Collins

Maths
in 5 minutes

Chief editor: Zhou Jieying
Consultant: Fan Lianghuo

CONTENTS

©HarperCollins*Publishers* 2019

HOW TO USE THIS BOOK

The best way to help your child to build their confidence in maths and improve their number skills is to give them lots and lots of practice in the key facts and skills.

Written by maths experts, this series will help your child to become fluent in number facts, and help them to recall them quickly – both are essential for succeeding in maths.

This book provides ready-to-practise questions that comprehensively cover the number curriculum for Year 6. It contains 40 topic-based tests, each 5 minutes long, to help your child build up their mathematical fluency day-by-day.

Each test is divided into three Steps:

- **Step 1: Warm-up (1 minute)**
  This exercise helps your child to revise maths they should already know and gives them preparation for Step 2.

- **Step 2: Rapid calculation ($2\frac{1}{2}$ minutes)**
  This exercise is a set of questions focused on the topic area being tested.

- **Step 3: Challenge ($1\frac{1}{2}$ minutes)**
  This is a more testing exercise designed to stretch your child's mental abilities.

Some of the tests also include:

- a Tip to help your child answer questions of a particular type.

- a Mind Gym puzzle – this is a further test of mental agility and is not included in the 5-minute time allocation.

Your child should attempt to answer as many questions as possible in the time allowed at each Step. Answers are provided at the back of the book.

To help to measure progress, each test includes boxes for recording the date of the test, the total score obtained and the total time taken.

ACKNOWLEDGEMENTS

The authors and publisher are grateful to the copyright holders for permission to use quoted materials and images.

All images are © HarperCollins*Publishers* Ltd and © Shutterstock.com

Every effort has been made to trace copyright holders and obtain their permission for the use of copyright material. The authors and publisher will gladly receive information enabling them to rectify any error or omission in subsequent editions. All facts are correct at time of going to press.

Published by Collins in association with East China Normal University Press

Published by Collins
An imprint of HarperCollins*Publishers*
1 London Bridge Street
London SE1 9GF

ISBN: 978-0-00-831113-1

First published 2019
This edition published 2020
Previously published by Letts

10 9 8 7 6 5 4 3

©HarperCollins*Publishers* Ltd. 2020,
©East China Normal University Press Ltd.,
©Zhou Jieying

British Library Cataloguing in Publication Data.

A CIP record of this book is available from the British Library.

Publisher: Fiona McGlade
Consultant: Fan Lianghuo
Authors: Hong Jinsong, Zhou Jieying and Chen Weihua
Editors: Ni Ming and Xu Huiping
Contributor: Paul Hodge
Project Management and Editorial: Richard Toms, Lauren Murray and Marie Taylor
Cover Design: Kevin Robbins and Sarah Duxbury
Inside Concept Design: Paul Oates and Ian Wrigley
Layout: Jouve India Private Limited
Printed in Great Britain by Bell and Bain Ltd, Glasgow

MIX
Paper from
responsible source
FSC C007454

This book is produced from independently certified FSC™ paper to ensure responsible forest management.

For more information visit:
www.harpercollins.co.uk/green

1 Knowing large numbers

Date: _____

Day of Week: _____

STEP 1 (1 min) Warm-up

Start the timer

Fill in the missing numbers. The first one has been done for you.

5 342 475 = | 5 000 000 | + | 300 000 | + | 40 000 | + | 2000 | + | 400 | + | 70 | + | 5 |

3 571 596 = ☐ + ☐ + ☐ + ☐ + ☐ + ☐ + ☐

9 436 034 = ☐ + ☐ + ☐ + ☐ + ☐ + ☐ + ☐

6 993 905 = ☐ + ☐ + ☐ + ☐ + ☐ + ☐ + ☐

2 285 439 = ☐ + ☐ + ☐ + ☐ + ☐ + ☐ + ☐

STEP 2 (2.5 min) Rapid calculation

Start the timer

Complete the table.

Description of number	Number in digits
3 millions, 4 hundred thousands, 2 ten thousands, 3 thousands, 7 hundreds and 4 ones	
7 millions, 2 hundred thousands, 9 ten thousands, 3 thousands, 8 hundreds, 6 tens and 8 ones	
9 millions, 3 hundred thousands, 7 ten thousands, 9 thousands and 2 tens	
5 millions, 2 ten thousands, 8 thousands, 3 hundreds and 9 ones	
4 millions, 6 hundred thousands, 3 thousands, 1 hundred and 8 ones	
1 million, 7 thousands and 4 ones	

STEP 3 (1.5 min) Challenge

Start the timer

Complete the table.

Description of number	Number in digits	Read as
9 millions and 378 thousands		
8 millions, 52 ten thousands and 19 ones		
68 hundred thousands, 7 thousands and 39 tens		

Time spent: _____ min _____ sec. Total: _____ out of 16

Rounding of large numbers

2

STEP 1 ⏱(1 min) Warm-up

TIP *Rounding: If the first of the digits to be removed is 4 or less, drop the digit and replace it and all the digits to its right with zero (rounding down).*

If the first of the digits to be removed is 5 or greater, replace it and all digits to its right with zero and increase the digit to the left by 1 (rounding up).

Round the numbers to the nearest ten thousand.

4 479 483 ≈ ☐   1 743 347 ≈ ☐   5 005 498 ≈ ☐

7 403 344 ≈ ☐   9 534 238 ≈ ☐   8 244 407 ≈ ☐

2 439 293 ≈ ☐   3 547 348 ≈ ☐

STEP 2 ⏱(2.5 min) Rapid calculation

Round the numbers to the nearest hundred thousand.

3 023 873 ≈ ☐   6 643 843 ≈ ☐   5 485 077 ≈ ☐

4 934 131 ≈ ☐   1 746 473 ≈ ☐   7 510 329 ≈ ☐

8 039 374 ≈ ☐   3 774 935 ≈ ☐   5 357 320 ≈ ☐

6 634 273 ≈ ☐   9 532 149 ≈ ☐   1 163 222 ≈ ☐

STEP 3 ⏱(1.5 min) Challenge

Round the numbers below to the nearest ten thousand, hundred thousand and million.

	Ten thousand	Hundred thousand	Million
4 495 054			
6 373 293			
2 346 523			
7 837 248			

Date: _____

Day of Week: _____

STEP 1 (1 min) Warm-up

Start the timer

Answer these.

```
    1 5 3 8          5 3 4 5          2 4 7 2          3 7 3 1
×         5        ×         4        ×         5        ×         6
_____          _____        _____         _____

_____          _____        _____         _____
```

STEP 2 (2.5 min) Rapid calculation

Start the timer

Answer these.

```
    3 4 4 3          2 1 1 9          3 4 1 2          4 3 7 8
×         2        ×         7        ×         4        ×         3
_____          _____        _____         _____

_____          _____        _____         _____


    2 3 4 3          3 2 8 4          2 4 3 4          4 5 3 7
×         4        ×         8        ×         9        ×         5
_____          _____        _____         _____

_____          _____        _____         _____
```

STEP 3 (1.5 min) Challenge

Start the timer

Answer these.

```
    2 3 4 2          4 5 6 4          6 0 2 4          9 4 5 6
×       6 2        ×       5 6        ×       3 7        ×       4 9
_____          _____        _____         _____
```

Time spent: _____ min _____ sec. Total: _____ out of 16

Dividing four-digit numbers by a one-digit number

4

STEP 1 (1 min) **Warm-up**

Start the timer

Work out these.

9 | 8 5 5   6 | 5 2 8   4 | 3 4 0

7 | 6 7 2   6 | 3 5 4   5 | 3 7 0

STEP 2 (2.5 min) **Rapid calculation**

Start the timer

Work out these.

2 | 7 0 4 8   7 | 3 0 5 9   7 | 7 9 8 0

3 | 1 4 7 9   2 | 2 8 2 2   4 | 9 6 5 2

6 | 2 5 6 8   5 | 5 8 5 0   8 | 6 8 4 8

STEP 3 (1.5 min) **Challenge**

Start the timer

Work out these.

8 | 1 8 9 8   r   5 | 7 7 0 7   r   7 | 4 4 5 0   r   4 | 2 5 9 5   r

8 | 6 0 8 2   r   6 | 2 8 6 6   r   7 | 9 7 2 0   r   6 | 6 3 6 0   r

---

Time spent: _____ min _____ sec. Total: _____ out of 23

5 Dividing four-digit numbers by a two-digit number

Date: _____

Day of Week: _____

Start the timer

Work out these.

14 | 3 6 4

21 | 3 5 7

33 | 6 2 7

25 | 6 2 5

17 | 5 1 0

24 | 8 1 6

STEP 2 (2.5 min) **Rapid calculation**

Start the timer

Work out these.

26 | 6 2 1 4

35 | 1 7 1 5

14 | 8 2 1 8

41 | 2 6 6 5

37 | 5 5 5 0

24 | 7 2 2 4

12 | 7 0 6 8

11 | 4 2 1 3

22 | 5 3 9 0

STEP 3 (1.5 min) **Challenge**

Start the timer

Work out these.

57 | 4 2 4 1   r

73 | 1 3 9 1   r

44 | 4 9 4 2   r

70 | 9 2 5 3   r

56 | 3 8 6 2   r

54 | 6 6 8 7   r

Time spent: _____ min _____ sec. Total: _____ out of 21

©HarperCollinsPublishers 2019

Date: _____

Day of Week: _____

STEP 1 (1 min) **Warm-up**

Start the timer

Answer these.

$8 \times 83 =$ ☐

$45 \times 11 =$ ☐

$360 + 253 =$ ☐

$470 - 380 =$ ☐

$848 \div 4 =$ ☐

$522 \div 3 =$ ☐

$505 - 80 =$ ☐

$1360 \div 8 =$ ☐

STEP 2 (2.5 min) **Rapid calculation**

Start the timer

Answer these.

$8 \times (69 + 14) =$ ☐

$45 \times 11 - 4 =$ ☐

$(245 + 277) \div 3 =$ ☐

$225 \times (44 \div 2) =$ ☐

$13 \times 2 \times 25 =$ ☐

$510 \div 6 - 1 =$ ☐

$90 - 60 \div 5 =$ ☐

$85 + 15 \times 40 =$ ☐

$(420 - 220) \times 38 =$ ☐

$950 \div (27 + 23) =$ ☐

$(2790 + 90) \div 30 =$ ☐

$46 \times 10 - 290 =$ ☐

$450 \times (72 - 67) =$ ☐

$2790 + 90 \div 3 =$ ☐

$50 \times 5 + 600 =$ ☐

STEP 3 (1.5 min) **Challenge**

Start the timer

Answer these.

$55 \div 11 \times 12 =$ ☐

$48 \times 51 - 48 =$ ☐

$312 \div 12 \times 9 =$ ☐

$4 \times 26 \div 13 + 2 =$ ☐

$45 \times 11 - (34 - 30) =$ ☐

$46 \times 10 - 290 \div 10 =$ ☐

$47 \times 10 - 76 \times 5 =$ ☐

$848 \div (3 \times 8 - 20) =$ ☐

$(420 + 3 \times 30) \div 6 =$ ☐

Date: _____

Day of Week: _____

STEP 1 (1 min) **Warm-up**

Start the timer

1. Find the perimeter (P) of each square where L is the side length.

   L = 2 cm, P = ☐ cm   L = 5 cm, P = ☐ cm   L = 17 cm, P = ☐ cm

2. Find the perimeter (P) of each rectangle where L is the length and W is the width.

   L = 3 cm, W = 5 cm, P = ☐ cm   L = 6 cm, W = 2 cm, P = ☐ cm

   L = 12 cm, W = 10 cm, P = ☐ cm

STEP 2 (2.5 min) **Rapid calculation**

Start the timer

1. Find the perimeter (P) of each square where L is the side length.

   L = 0.3 cm, P = ☐ cm   L = 0.6 cm, P = ☐ cm   L = 1.4 cm, P = ☐ cm

   L = 4.5 cm, P = ☐ cm   L = 2.9 m, P = ☐ m   L = 0.14 m, P = ☐ m

   L = 16 m, P = ☐ m   L = 11.2 m, P = ☐ m

2. Find the perimeter (P) of each rectangle where L is the length and W is the width.

   L = 2.1 m, W = 1.3 m, P = ☐ m   L = 0.2 m, W = 1.8 m, P = ☐ m

   L = 1.5 m, W = 2.5 m, P = ☐ m   L = 3.6 m, W = 4.6 m, P = ☐ m

STEP 3 (1.5 min) **Challenge**

Start the timer

1. Find the side length (L) of each square from its perimeter (P).

   P = 3.2 cm, L = ☐ cm   P = 4.4 cm, L = ☐ cm   P = 0.16 m, L = ☐ m

2. Find the length of the unknown side in each rectangle.

   (P = perimeter, L = length and W = width)

   P = 2.8 m, L = 0.6 m, W = ☐ m   P = 4.4 m, W = 1.2 m, L = ☐ m

   P = 12.6 cm, L = 2.4 cm, W = ☐ cm

Time spent: _____ min _____ sec. Total: _____ out of 24   ©HarperCollinsPublishers 2019

Start the timer

STEP 1 (1 min) Warm-up

Find the area (A) of the shapes with the given dimensions.

L is the (side or base) length, W is the width and H is the height.

Rectangle: $L = 6\,m$, $W = 8\,m$, $A = \boxed{}\,m^2$

Rectangle: $L = 8\,cm$, $W = 12\,cm$, $A = \boxed{}\,cm^2$

Square: $L = 5\,cm$, $A = \boxed{}\,cm^2$

Parallelogram: $L = 2\,cm$, $H = 3\,cm$, $A = \boxed{}\,cm^2$

Triangle: $L = 4\,m$, $H = 16\,m$, $A = \boxed{}\,m^2$

Triangle: $L = 3\,m$, $H = 4\,m$, $A = \boxed{}\,m^2$

STEP 2 (2.5 min) Rapid calculation

Start the timer

Use the information given about each shape to work out the unknown values.

Rectangle: $L = 0.7\,cm$, $W = 6\,cm$, $A = \boxed{}\,cm^2$

Rectangle: $A = 3.6\,m^2$, $L = 2\,m$, $W = \boxed{}\,m$

Square: $L = 0.3\,cm$, $A = \boxed{}\,cm^2$

Rectangle: $L = 2\,cm$, $W = 0.3\,cm$, $A = \boxed{}\,cm^2$

Parallelogram: $A = 0.8\,m^2$, $L = 0.5\,m$, $H = \boxed{}\,m$

Parallelogram: $A = 7.8\,cm^2$, $L = 3\,cm$, $H = \boxed{}\,cm$

Parallelogram: $A = 4.5\,m^2$, $L = 0.3\,m$, $H = \boxed{}\,m$

Triangle: $L = 5.6\,cm$, $H = 6\,cm$, $A = \boxed{}\,cm^2$

Triangle: $A = 17\,cm^2$, $L = 5\,cm$, $H = \boxed{}\,cm$

Triangle: $A = 10\,m^2$, $H = 2\,m$, $L = \boxed{}\,m$

STEP 3 (1.5 min) Challenge

Start the timer

Use the information given about each shape to work out the unknown values. Take care with the units.

Rectangle: $L = 4\,cm$, $W = 25\,cm$, $A = \boxed{}\,m^2$

Rectangle: $A = 1.35\,m^2$, $W = 10\,cm$, $L = \boxed{}\,cm$

Square: $L = 1.1\,m$, $A = \boxed{}\,cm^2$

Parallelogram: $A = 20\,cm^2$, $L = 0.05\,m$, $H = \boxed{}\,cm$

Triangle: $A = 4\,m^2$, $H = 40\,cm$, $L = \boxed{}\,cm$

Triangle: $A = 3.4\,cm^2$, $H = 0.04\,m$, $L = \boxed{}\,cm$

©HarperCollinsPublishers 2019

Time spent: _____ min _____ sec. Total: _____ out of 22

11

9 Fractions and division

Date: _____

Day of Week: _____

STEP 1 (1 min) Warm-up

Start the timer

Fill in the missing numbers.

$5 \div 13 = \dfrac{\Box}{\Box}$

$\Box \div 8 = \dfrac{5}{8}$

$4 \div 7 = \dfrac{\Box}{\Box}$

$\dfrac{11}{6} = \Box \div \Box$

$\dfrac{7}{3} = \Box \div \Box$

$6 \div 17 = \dfrac{\Box}{\Box}$

$\Box \div 9 = \dfrac{2}{9}$

$7 \div 15 = \dfrac{\Box}{\Box}$

STEP 2 (2.5 min) Rapid calculation

Start the timer

Fill in the missing numbers.

$3 \div 11 = \dfrac{\Box}{\Box}$

$\Box \div 3 = \dfrac{8}{3}$

$14 \div 17 = \dfrac{\Box}{\Box}$

$\dfrac{7}{12} = \Box \div \Box$

$\dfrac{36}{13} = \Box \div \Box$

$2 \div 5 = \dfrac{\Box}{\Box}$

$\Box \div 19 = \dfrac{2}{19}$

$27 \div 25 = \dfrac{\Box}{\Box}$

$15 \div 7 = \dfrac{\Box}{\Box}$

$4 \div \Box = \dfrac{4}{15}$

$17 \div 6 = \dfrac{\Box}{\Box}$

$\dfrac{9}{2} = \Box \div \Box$

$\dfrac{4}{7} = \Box \div \Box$

$4 \div 18 = \dfrac{\Box}{\Box}$

$\Box \div 11 = \dfrac{8}{11}$

$3 \div 11 = \dfrac{\Box}{\Box}$

$5 \div \Box = \dfrac{5}{18}$

$\dfrac{12}{5} = \Box \div \Box$

$10 \div \Box = \dfrac{10}{15}$

$\dfrac{20}{3} = \Box \div \Box$

STEP 3 (1.5 min) Challenge

Start the timer

Fill in the missing numbers.

$37\,\text{min} = \dfrac{\Box}{\Box}\,\text{h}$

$19\,\text{g} = \dfrac{\Box}{\Box}\,\text{kg}$

$3\,\text{cm} = \dfrac{\Box}{\Box}\,\text{m}$

$27\,\text{h} = \dfrac{\Box}{\Box}\,\text{days}$

$53\,\text{mL} = \dfrac{\Box}{\Box}\,\text{L}$

$327\,\text{g} = \dfrac{\Box}{\Box}\,\text{kg}$

$17\,\text{cm} = \dfrac{\Box}{\Box}\,\text{m}$

$50\,\text{m} = \dfrac{\Box}{\Box}\,\text{km}$

Time spent: _____ min _____ sec. Total: _____ out of 36

STEP 1 (1 min) Warm-up

Start the timer

TIP *Use your knowledge of multiples to help find equivalent fractions.*

Complete these equivalent fractions.

$\dfrac{2}{3} = \dfrac{\boxed{}}{6}$ $\dfrac{10}{15} = \dfrac{\boxed{}}{3}$ $\dfrac{1}{4} = \dfrac{5}{\boxed{}}$ $\dfrac{6}{18} = \dfrac{2}{\boxed{}}$

$\dfrac{5}{6} = \dfrac{\boxed{}}{12}$ $\dfrac{\boxed{}}{7} = \dfrac{24}{42}$ $\dfrac{5}{8} = \dfrac{20}{\boxed{}}$ $\dfrac{4}{\boxed{}} = \dfrac{48}{60}$

STEP 2 (2.5 min) Rapid calculation

Start the timer

Complete these equivalent fractions.

$\dfrac{2}{7} = \dfrac{\boxed{}}{21}$ $\dfrac{10}{25} = \dfrac{\boxed{}}{5}$ $\dfrac{1}{6} = \dfrac{4}{\boxed{}}$ $\dfrac{6}{12} = \dfrac{3}{\boxed{}}$ $\dfrac{20}{24} = \dfrac{\boxed{}}{6}$

$\dfrac{\boxed{}}{14} = \dfrac{24}{42}$ $\dfrac{3}{4} = \dfrac{\boxed{}}{16}$ $\dfrac{4}{\boxed{}} = \dfrac{24}{54}$ $\dfrac{5}{6} = \dfrac{30}{\boxed{}}$ $\dfrac{20}{30} = \dfrac{\boxed{}}{6}$

$\dfrac{1}{3} = \dfrac{5}{\boxed{}}$ $\dfrac{7}{9} = \dfrac{\boxed{}}{45}$ $\dfrac{6}{24} = \dfrac{2}{\boxed{}}$ $\dfrac{5}{7} = \dfrac{30}{\boxed{}}$ $\dfrac{\boxed{}}{6} = \dfrac{28}{42}$

STEP 3 (1.5 min) Challenge

Start the timer

Complete these equivalent fractions.

$\dfrac{1}{6} = \dfrac{\boxed{}}{12}$ $\dfrac{15}{30} = \dfrac{\boxed{}}{4}$ $\dfrac{2}{4} = \dfrac{5}{\boxed{}}$ $\dfrac{6}{21} = \dfrac{2}{\boxed{}}$

$\dfrac{5}{6} = \dfrac{\boxed{}}{72}$ $\dfrac{\boxed{}}{100} = \dfrac{3}{4}$ $\dfrac{2}{8} = \dfrac{5}{\boxed{}}$ $\dfrac{4}{\boxed{}} = \dfrac{16}{20}$

Time spent: _____ min _____ sec. Total: _____ out of 31

Date: _____

Day of Week: _____

STEP 1 (1 min) Warm-up

Start the timer

Fill in the missing numbers.

$$\frac{\boxed{}}{24} = \frac{16}{\boxed{}} = \frac{1}{3} = \boxed{} \div \boxed{}$$

$$\frac{24}{\boxed{}} = \frac{\boxed{}}{81} = \frac{2}{9} = 12 \div \boxed{}$$

$$\frac{15}{\boxed{}} = \frac{\boxed{}}{48} = 0.125 = \boxed{1} \div \boxed{}$$

$$\frac{\boxed{}}{25} = \frac{32}{\boxed{}} = 0.8 = \boxed{24} \div \boxed{}$$

STEP 2 (2.5 min) Rapid calculation

Start the timer

Fully simplify the fractions.

$$\frac{4}{8} = \boxed{}$$

$$\frac{2}{8} = \boxed{}$$

$$\frac{24}{54} = \boxed{}$$

$$\frac{56}{96} = \boxed{}$$

$$\frac{6}{9} = \boxed{}$$

$$\frac{12}{30} = \boxed{}$$

$$\frac{4}{22} = \boxed{}$$

$$\frac{18}{42} = \boxed{}$$

$$\frac{3}{12} = \boxed{}$$

$$\frac{8}{18} = \boxed{}$$

$$\frac{39}{65} = \boxed{}$$

$$\frac{24}{36} = \boxed{}$$

STEP 3 (1.5 min) Challenge

Start the timer

Fully simplify the fractions.

$$\frac{16}{84} = \boxed{}$$

$$\frac{49}{203} = \boxed{}$$

$$\frac{45}{99} = \boxed{}$$

$$\frac{74}{111} = \boxed{}$$

$$\frac{14}{91} = \boxed{}$$

$$\frac{15}{125} = \boxed{}$$

$$\frac{40}{95} = \boxed{}$$

$$\frac{96}{128} = \boxed{}$$

Time spent: _____ min _____ sec. Total: _____ out of 33

STEP 1 $\binom{1}{min}$ Warm-up

Start the timer

TIP *To decide whether a fraction is less than $\frac{1}{2}$, test whether the denominator is greater than twice the numerator. If it is, the fraction is less than $\frac{1}{2}$, e.g. $\frac{3}{7}$; if it is not, it is greater than $\frac{1}{2}$, e.g. $\frac{3}{5}$.*

Fill in the boxes with **>, <** or **=**.

$3 \; \square \; \frac{12}{4}$ $\frac{4}{5} \; \square \; \frac{5}{6}$ $\frac{3}{8} \; \square \; \frac{5}{16}$ $1.5 \; \square \; \frac{3}{4}$

$\frac{1}{2} \; \square \; 0.5$ $\frac{10}{3} \; \square \; \frac{13}{4}$ $\frac{2}{3} \; \square \; \frac{2}{5}$ $\frac{3}{20} \; \square \; \frac{1}{5}$

STEP 2 $\binom{2.5}{min}$ **Rapid calculation**

Start the timer

Fill in the boxes with **>, <** or **=**.

$\frac{1}{3} \; \square \; \frac{1}{4}$ $\frac{4}{5} \; \square \; \frac{4}{7}$ $\frac{3}{8} \; \square \; \frac{5}{8}$ $\frac{9}{16} \; \square \; \frac{9}{25}$

$\frac{1}{2} \; \square \; \frac{2}{3}$ $\frac{4}{9} \; \square \; \frac{3}{7}$ $\frac{3}{10} \; \square \; \frac{5}{12}$ $\frac{9}{11} \; \square \; \frac{7}{12}$

$\frac{8}{9} \; \square \; \frac{5}{6}$ $\frac{1}{3} \; \square \; \frac{2}{7}$ $\frac{2}{5} \; \square \; \frac{3}{10}$ $\frac{5}{8} \; \square \; \frac{7}{10}$

$\frac{7}{8} \; \square \; \frac{4}{5}$ $\frac{2}{5} \; \square \; \frac{3}{8}$ $\frac{3}{7} \; \square \; \frac{5}{12}$ $\frac{9}{10} \; \square \; \frac{19}{25}$

STEP 3 $\binom{1.5}{min}$ **Challenge**

Start the timer

Look at the fractions in the box.

| $\frac{2}{5}$ | $\frac{1}{6}$ | $\frac{3}{8}$ | $\frac{4}{9}$ | $\frac{6}{7}$ | $\frac{6}{11}$ | $\frac{7}{10}$ | $\frac{3}{7}$ | $\frac{3}{5}$ | $\frac{2}{3}$ | $\frac{3}{4}$ | $\frac{5}{12}$ |

1. Write down the fractions that are less than $\frac{1}{2}$.

....

2. Write down the fractions that are greater than $\frac{1}{2}$.

....

Date: _____

Day of Week: _____

STEP 1 (1 min) Warm-up

Start the timer

Answer these. Use improper fractions where appropriate.

$\frac{23}{50} + \frac{2}{50} = \boxed{}$ $\frac{16}{9} - \frac{2}{9} = \boxed{}$ $\frac{1}{8} + \frac{15}{8} = \boxed{}$ $\frac{17}{18} - \frac{7}{18} = \boxed{}$

$\frac{3}{16} + \frac{5}{16} = \boxed{}$ $\frac{53}{78} - \frac{25}{78} = \boxed{}$ $\frac{13}{36} + \frac{7}{36} = \boxed{}$ $\frac{17}{8} - \frac{11}{8} = \boxed{}$

STEP 2 (2.5 min) Rapid calculation

Start the timer

Answer these. Use improper fractions where appropriate.

$\frac{1}{2} + \frac{3}{2} = \boxed{}$ $\frac{7}{3} - \frac{1}{3} = \boxed{}$ $\frac{11}{12} + \frac{5}{12} = \boxed{}$ $\frac{8}{15} - \frac{3}{15} = \boxed{}$

$\frac{2}{9} + \frac{1}{9} = \boxed{}$ $\frac{7}{25} - \frac{2}{25} = \boxed{}$ $\frac{11}{10} + \frac{7}{10} = \boxed{}$ $1 - \frac{7}{9} = \boxed{}$

$\frac{1}{20} + \frac{4}{20} = \boxed{}$ $\frac{3}{20} + \frac{7}{20} = \boxed{}$ $\frac{9}{70} - \frac{1}{70} = \boxed{}$ $\frac{2}{15} + \frac{7}{15} = \boxed{}$

$\frac{7}{10} - \frac{1}{10} = \boxed{}$ $\frac{4}{15} + \frac{8}{15} = \boxed{}$ $\frac{23}{70} - \frac{3}{70} = \boxed{}$ $\frac{1}{20} + \frac{9}{20} = \boxed{}$

$\frac{7}{15} - \frac{1}{15} = \boxed{}$ $\frac{5}{4} + \frac{1}{4} = \boxed{}$ $\frac{9}{17} - \frac{3}{17} = \boxed{}$ $\frac{9}{8} + \frac{3}{8} = \boxed{}$

STEP 3 (1.5 min) Challenge

Start the timer

Answer these. Use improper fractions where appropriate.

$\frac{1}{18} + \frac{3}{18} + \frac{7}{18} = \boxed{}$ $\frac{17}{12} - \frac{1}{12} - \frac{5}{12} = \boxed{}$

$\frac{1}{24} + \frac{3}{24} + \frac{5}{24} = \boxed{}$ $\frac{17}{10} - \frac{3}{10} - \frac{9}{10} = \boxed{}$

$\frac{1}{13} + \frac{5}{13} + \frac{7}{13} = \boxed{}$ $\frac{13}{8} - \frac{1}{8} - \frac{5}{8} = \boxed{}$

$\frac{1}{12} + \frac{5}{12} + \frac{7}{12} = \boxed{}$ $\frac{7}{10} - \frac{1}{10} - \frac{1}{10} = \boxed{}$

Time spent: _____ min _____ sec. Total: _____ out of 36

Date: _____

Day of Week: _____

STEP 1 (1 min) Warm-up

Start the timer

Answer these. Use improper fractions where appropriate.

$\frac{2}{3} + \frac{2}{5} = \boxed{}$ $\frac{6}{7} - \frac{2}{5} = \boxed{}$ $\frac{2}{7} + \frac{5}{9} = \boxed{}$ $\frac{5}{6} - \frac{1}{3} = \boxed{}$

$\frac{2}{5} + \frac{5}{7} = \boxed{}$ $\frac{7}{10} - \frac{1}{5} = \boxed{}$ $\frac{2}{7} + \frac{9}{11} = \boxed{}$ $\frac{7}{9} - \frac{1}{4} = \boxed{}$

STEP 2 (2.5 min) Rapid calculation

Start the timer

Answer these. Use improper fractions where appropriate.

$\frac{1}{2} + \frac{3}{8} = \boxed{}$ $\frac{7}{12} - \frac{1}{3} = \boxed{}$ $\frac{2}{3} + \frac{5}{7} = \boxed{}$ $\frac{5}{8} - \frac{1}{4} = \boxed{}$

$\frac{2}{7} + \frac{1}{3} = \boxed{}$ $\frac{7}{8} - \frac{1}{2} = \boxed{}$ $\frac{1}{5} + \frac{9}{10} = \boxed{}$ $\frac{1}{2} - \frac{1}{9} = \boxed{}$

$\frac{2}{5} + \frac{4}{9} = \boxed{}$ $\frac{3}{8} + \frac{1}{3} = \boxed{}$ $\frac{9}{12} - \frac{1}{2} = \boxed{}$ $\frac{2}{9} + \frac{1}{3} = \boxed{}$

$\frac{3}{5} - \frac{1}{10} = \boxed{}$ $\frac{3}{4} + \frac{5}{6} = \boxed{}$ $\frac{5}{8} - \frac{2}{9} = \boxed{}$ $\frac{1}{3} + \frac{3}{4} = \boxed{}$

$\frac{7}{12} - \frac{1}{6} = \boxed{}$ $\frac{3}{5} + \frac{2}{7} = \boxed{}$ $\frac{13}{18} - \frac{2}{9} = \boxed{}$ $\frac{2}{5} + \frac{3}{7} = \boxed{}$

STEP 3 (1.5 min) Challenge

Start the timer

Answer these. Use improper fractions where appropriate.

$\frac{1}{8} + \frac{3}{4} + \frac{1}{2} = \boxed{}$ $\frac{7}{12} - \frac{1}{3} - \frac{1}{12} = \boxed{}$

$\frac{1}{14} + \frac{3}{7} + \frac{5}{14} = \boxed{}$ $\frac{17}{20} - \frac{3}{10} - \frac{9}{20} = \boxed{}$

$\frac{1}{9} + \frac{5}{6} + \frac{1}{3} = \boxed{}$ $\frac{13}{14} - \frac{1}{14} - \frac{5}{7} = \boxed{}$

$\frac{1}{18} + \frac{5}{9} + \frac{1}{3} = \boxed{}$ $\frac{7}{15} - \frac{1}{15} - \frac{1}{5} = \boxed{}$

Date: _____

Day of Week: _____

STEP 1 ⏱ (1 min) **Warm-up**

Start the timer 👆

Answer these.

$\dfrac{1}{2} \times \dfrac{1}{3} =$ ☐

$\dfrac{1}{4} \times \dfrac{1}{5} =$ ☐

$\dfrac{3}{7} \times \dfrac{3}{4} =$ ☐

$\dfrac{5}{12} \times \dfrac{1}{3} =$ ☐

$\dfrac{3}{5} \times \dfrac{1}{8} =$ ☐

$\dfrac{5}{11} \times \dfrac{2}{3} =$ ☐

$\dfrac{5}{14} \times \dfrac{1}{3} =$ ☐

$\dfrac{1}{3} \times 84 =$ ☐

$64 \times \dfrac{1}{4} =$ ☐

STEP 2 ⏱ (2.5 min) **Rapid calculation**

Start the timer 👆

Answer these.

$\dfrac{7}{15} \times \dfrac{2}{3} =$ ☐

$\dfrac{3}{8} \times \dfrac{5}{24} =$ ☐

$\dfrac{16}{63} \times \dfrac{21}{26} =$ ☐

$4 \times \dfrac{5}{24} =$ ☐

$\dfrac{13}{38} \times \dfrac{19}{39} =$ ☐

$\dfrac{7}{12} \times 48 =$ ☐

$\dfrac{11}{36} \times \dfrac{24}{55} =$ ☐

$\dfrac{2}{3} \times \dfrac{9}{14} =$ ☐

$\dfrac{5}{13} \times 52 =$ ☐

$\dfrac{13}{18} \times \dfrac{15}{26} =$ ☐

$68 \times \dfrac{11}{17} =$ ☐

$\dfrac{17}{90} \times \dfrac{15}{34} =$ ☐

STEP 3 ⏱ (1.5 min) **Challenge**

Start the timer 👆

Answer these. Use mixed numbers where appropriate.

$\dfrac{7}{15} \times \dfrac{10}{49} \times \dfrac{3}{5} =$ ☐

$\dfrac{5}{42} \times \dfrac{7}{24} \times 48 =$ ☐

$\dfrac{10}{13} \times 52 \times \dfrac{3}{120} =$ ☐

$\dfrac{11}{36} \times \dfrac{24}{55} \times \dfrac{5}{18} =$ ☐

$\dfrac{8}{27} \times \dfrac{9}{14} \times \dfrac{21}{36} =$ ☐

$4 \times \dfrac{25}{44} \times \dfrac{22}{75} =$ ☐

$\dfrac{15}{28} \times \dfrac{7}{10} \times \dfrac{8}{16} =$ ☐

$\dfrac{13}{54} \times \dfrac{18}{49} \times \dfrac{21}{26} =$ ☐

Time spent: _____ min _____ sec. Total: _____ out of 29

Date: _____

Day of Week: _____

STEP 1 (1 min) Warm-up

Start the timer

Fill in the missing numbers.

There are ☐ $\frac{1}{12}$ s in $\frac{7}{12}$.

There are ☐ $\frac{1}{9}$ s in $\frac{4}{9}$.

There are ☐ $\frac{1}{24}$ s in 1.

☐ $\frac{1}{7}$ s makes $\frac{3}{7}$.

☐ $\frac{1}{18}$ s makes $\frac{5}{18}$.

☐ $\frac{1}{45}$ s makes $\frac{14}{45}$.

There are ☐ $\frac{1}{21}$ s in $\frac{8}{21}$.

☐ $\frac{1}{6}$ s makes 1.

There are ☐ $\frac{1}{32}$ s in $\frac{19}{32}$.

STEP 2 (2.5 min) Rapid calculation

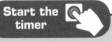
Start the timer

Write each answer as a fraction or a mixed number.

$17 \div 18 =$ ☐

$7 \div 15 =$ ☐

$11 \div 19 =$ ☐

$24 \div 37 =$ ☐

$6 \div 11 =$ ☐

$18 \div 7 =$ ☐

$25 \div 12 =$ ☐

$23 \div 5 =$ ☐

$38 \div 9 =$ ☐

$13 \div 28 =$ ☐

$28 \div 13 =$ ☐

$95 \div 39 =$ ☐

$75 \div 32 =$ ☐

$29 \div 35 =$ ☐

$41 \div 9 =$ ☐

STEP 3 (1.5 min) Challenge

Start the timer

1. Fill in the missing numbers.

$\frac{23}{21} =$ ☐ $\div$ ☐

$\frac{42}{95} =$ ☐ $\div$ ☐

$\frac{5}{18} =$ ☐ $\div$ ☐

$\frac{7}{26} =$ ☐ $\div$ ☐

2. Give your answers to these as a mixed number.

$95 \div 45 =$ ☐

$32 \div 12 =$ ☐

$49 \div 13 =$ ☐

$96 \div 44 =$ ☐

$80 \div 6 =$ ☐

$76 \div 14 =$ ☐

Time spent: _____ min _____ sec. Total: _____ out of 34

Date: _____

Day of Week: _____

STEP 1 (1 min) Warm-up

Start the timer

Answer these.

$\frac{1}{3} \times \frac{1}{3} =$ ☐

$\frac{5}{18} \times \frac{1}{5} =$ ☐

$\frac{4}{9} \times \frac{1}{4} =$ ☐

$\frac{1}{8} \times \frac{2}{7} =$ ☐

$\frac{3}{5} \times \frac{10}{33} =$ ☐

$\frac{5}{8} \times \frac{2}{5} =$ ☐

$\frac{5}{6} \times \frac{4}{15} =$ ☐

$\frac{3}{14} \times \frac{7}{9} =$ ☐

$\frac{1}{4} \times \frac{2}{5} =$ ☐

STEP 2 (2.5 min) Rapid calculation

Start the timer

TIP *Dividing a fraction by a whole number is equivalent to multiplying the fraction by the reciprocal of the divisor.*

For example, $\frac{15}{28} \div 6 = \frac{15}{28} \times \frac{1}{6} = \frac{5}{56}$

Answer these.

$\frac{4}{9} \div 2 =$ ☐

$\frac{3}{7} \div 3 =$ ☐

$\frac{7}{18} \div 21 =$ ☐

$\frac{15}{16} \div 12 =$ ☐

$\frac{8}{9} \div 4 =$ ☐

$\frac{45}{56} \div 15 =$ ☐

$\frac{8}{13} \div 4 =$ ☐

$\frac{1}{3} \div 3 =$ ☐

$\frac{9}{14} \div 6 =$ ☐

$\frac{12}{23} \div 12 =$ ☐

$\frac{1}{3} \div 2 =$ ☐

$\frac{9}{26} \div 3 =$ ☐

STEP 3 (1.5 min) Challenge

Start the timer

Answer these.

$\frac{46}{91} \div 23 =$ ☐

$\frac{23}{90} \div 69 =$ ☐

$\frac{51}{110} \div 34 =$ ☐

$\frac{16}{49} \div 24 =$ ☐

$\frac{91}{105} \div 39 =$ ☐

$\frac{35}{120} \div 105 =$ ☐

$\frac{21}{80} \div 105 =$ ☐

$\frac{48}{95} \div 8 =$ ☐

Time spent: _____ min _____ sec. Total: _____ out of 29

Date: _____

Day of Week: _____

STEP 1 (1 min) **Warm-up**

Start the timer

Answer these.

7.6 − 4 = ☐

11 − 2.5 = ☐

2.5 + 0.5 = ☐

9 − 1.2 = ☐

14 − 4.5 = ☐

2.01 + 0.09 = ☐

7 − 3.5 = ☐

34 + 0.15 = ☐

STEP 2 (2.5 min) **Rapid calculation**

Start the timer

Fill in the missing numbers.

8 − ☐ = 3.5

11 − ☐ = 3.5

☐ − 4.5 = 5.5

2.5 − ☐ = 2

☐ − 3.6 = 2.8

2.7 − ☐ = 0.9

9 − ☐ = 1.3

☐ − 3.9 = 5.3

☐ − 4.7 = 3.9

8 − ☐ = 1.6

☐ − 6.3 = 5.6

☐ − 0.09 = 2.1

6 − ☐ = 1.2

☐ − 0.76 = 0.24

☐ − 0.26 = 0.18

8 + ☐ = 12.8

4.3 + ☐ = 8.7

☐ + 2.4 = 10.2

STEP 3 (1.5 min) **Challenge**

Start the timer

Answer these.

0.88 − 0.21 = ☐

0.85 + 0.25 = ☐

8.5 + 5 = ☐

0.86 − 0.24 = ☐

0.19 + 0.27 = ☐

7.7 + 8 + 2.3 = ☐

27 − 4.1 − 5.9 = ☐

0.4 + 0.8 + 0.6 = ☐

8 + 2.8 + 7.2 = ☐

©HarperCollinsPublishers 2019

Time spent: _____ min _____ sec. Total: _____ out of 35

21

Date: _____

Day of Week: _____

STEP 1 (1 min) Warm-up

Start the timer

Answer these.

$6 - 0.6 =$ ☐ $0.75 + 2.5 =$ ☐ $7.77 + 2.3 =$ ☐ $6.8 - 5 =$ ☐

$80 - 7.054 =$ ☐ $1.1 + 1.11 =$ ☐ $1.3 + 0.7 =$ ☐ $100 - 9.685 =$ ☐

STEP 2 (2.5 min) Rapid calculation

Start the timer

Answer these.

$3.88 + 10.2 =$ ☐ $25 - 6.25 =$ ☐ $89.98 - 0.89 =$ ☐

$0.45 + 9.55 =$ ☐ $19 - 0.9 =$ ☐ $4.1 + 1.12 =$ ☐

$17.8 + 9.5 =$ ☐ $4.4 + 2.2 =$ ☐ $100 - 55.45 =$ ☐

$11.47 - 0.74 =$ ☐ $2.2 + 16.8 + 2 =$ ☐ $72.5 + 16.5 + 7.5 =$ ☐

$1.5 + 6.3 + 9.4 =$ ☐ $34.82 + 4.18 - 20 =$ ☐ $18.3 + 5.5 + 1.7 =$ ☐

$45.6 + 21.2 - 5.3 =$ ☐ $3.24 + 5.24 + 1.3 =$ ☐ $10.5 - 4.25 - 1.75 =$ ☐

STEP 3 (1.5 min) Challenge

Start the timer

 TIP *1 metric tonne (t) = 1000 kg*

Convert these measures.

$0.69 t =$ ☐ kg $9 t \, 40 kg =$ ☐ kg $5 t - 600 kg =$ ☐ kg

$0.49 kg =$ ☐ g $48 000 kg =$ ☐ t $18 m - 16 cm =$ ☐ cm

$75 m =$ ☐ cm $180 mm =$ ☐ cm $780 mm + 250 mm =$ ☐ m

$4600 g =$ ☐ kg $96 m \, 8 cm =$ ☐ cm $7560 m =$ ☐ km ☐ m

Time spent: _____ min _____ sec. Total: _____ out of 38

©HarperCollins*Publishers* 2019

Date: _____

Day of Week: _____

STEP 1 ⏱(1 min) Warm-up

Start the timer

 TIP *To multiply a decimal number by a whole number mentally, first treat the decimal as a whole number by ignoring the decimal point and find the product using multiplication facts. Then count the number of decimal places and move the digits of the product the appropriate number of place values to the right, removing any zeroes at the end. For example:*

0.5 × 15:

5 × 15 = 75

Since 0.5 has one decimal place, move the digits in 75 one place to the right.

0.5 × 15 = 7.5

0.005 × 4:

5 × 4 = 20

Since 0.005 has three decimal places, move the digits in 20 three places to the right and remove the zero at the end.

0.005 × 4 = 0.02

Answer these.

$4 \times 5 = \boxed{}$ $0.5 \times 8 = \boxed{}$ $0.7 \times 20 = \boxed{}$ $0.9 \times 6 = \boxed{}$

$0.8 \times 12 = \boxed{}$ $0.4 \times 5 = \boxed{}$ $15 \times 0.3 = \boxed{}$ $0.25 \times 8 = \boxed{}$

STEP 2 ⏱(2.5 min) Rapid calculation

Start the timer

1. Round these numbers to two decimal places.

$94.8524 \approx \boxed{}$ $16.4912 \approx \boxed{}$ $51.4216 \approx \boxed{}$ $27.6411 \approx \boxed{}$

$36.4711 \approx \boxed{}$ $88.1234 \approx \boxed{}$ $61.3471 \approx \boxed{}$ $9.2561 \approx \boxed{}$

2. Round these numbers to one decimal place.

$51.0416 \approx \boxed{}$ $18.0124 \approx \boxed{}$ $84.6233 \approx \boxed{}$ $17.1697 \approx \boxed{}$

$87.6123 \approx \boxed{}$ $71.2147 \approx \boxed{}$ $85.2641 \approx \boxed{}$ $13.2817 \approx \boxed{}$

STEP 3 ⏱(1.5 min) Challenge

Start the timer

Answer these.

$75 \times 0.02 = \boxed{}$ $0.005 \times 6 = \boxed{}$ $0.0015 \times 8 = \boxed{}$ $0.05 \times 3 = \boxed{}$

$0.08 \times 4 = \boxed{}$ $0.007 \times 4 = \boxed{}$ $25 \times 0.04 = \boxed{}$ $12.5 \times 8 = \boxed{}$

Addition, subtraction and multiplication with decimals

Date: _____

Day of Week: _____

STEP 1 (1 min) **Warm-up**

Start the timer

Answer these.

$0.16 \times 5 =$ ☐ $0.45 \times 4 =$ ☐ $0.51 \times 2 =$ ☐ $0.42 \times 3 =$ ☐

$0.32 \times 6 =$ ☐ $0.81 \times 6 =$ ☐ $0.72 \times 6 =$ ☐ $0.46 \times 5 =$ ☐

STEP 2 (2.5 min) **Rapid calculation**

Start the timer

Answer these.

$3 - (3 \times 0.3) =$ ☐ $0.6 - (0.3 \times 2) =$ ☐ $0.8 + (0.5 \times 0.8) =$ ☐

$1 - (0.2 \times 4) =$ ☐ $2 + 0.25 \times 0.4 =$ ☐ $1.6 - 0.15 \times 8 =$ ☐

$0.3 \times 5 + 0.3 =$ ☐ $0.3 + 0.3 \times 0.9 =$ ☐ $0.6 - 0.6 \times 0.5 =$ ☐

$0.5 + 0.5 \times 3 =$ ☐ $0.9 \times 3 - 0.7 =$ ☐ $0.8 + 2 \times 0.21 =$ ☐

$4 \times 0.7 - 0.7 =$ ☐ $0.6 \times 0.5 + 0.7 =$ ☐ $0.6 \times 0.25 + 0.5 =$ ☐

STEP 3 (1.5 min) **Challenge**

Start the timer

Answer these.

$0.75 \times 2 - 0.1 =$ ☐ $2.5 - 0.95 \times 2 =$ ☐ $0.45 \times 0.4 + 0.72 =$ ☐

$0.5 - 0.5 \times 0.8 =$ ☐ $4.62 - 0.62 \times 5 =$ ☐ $0.5 \times 0.2 + 1.9 =$ ☐

$0.1 + 0.33 \times 0.3 =$ ☐ $0.25 \times 0.8 + 3.8 =$ ☐ $2.1 + 0.95 \times 2 =$ ☐

Time spent: _____ min _____ sec. Total: _____ out of 32

STEP 1 (1 min) **Warm-up**

Start the timer

TIP *The methods for dividing decimals mentally are similar to those used for whole numbers. The main difference is that the decimals place of the quotient (answer) must be in line with the dividend (the number being divided). Always look for a smart way to make calculation easier.*

For example, to work out 42.7 ÷ 7, separate 42.7 into 42 + 0.7.

$42.7 ÷ 7 = (42 + 0.7) ÷ 7 = 42 ÷ 7 + 0.7 ÷ 7 = 6 + 0.1 = 6.1$

Answer these.

6.6 ÷ 6 = ☐   8.4 ÷ 2 = ☐   3.9 ÷ 3 = ☐   4.6 ÷ 2 = ☐

6.9 ÷ 3 = ☐   1.2 ÷ 3 = ☐   5.4 ÷ 3 = ☐   2.4 ÷ 3 = ☐

STEP 2 (2.5 min) **Rapid calculation**

Start the timer

Answer these.

0.16 ÷ 4 = ☐   0.04 ÷ 2 = ☐   0.09 ÷ 3 = ☐

7.5 ÷ 3 = ☐   9.6 ÷ 6 = ☐   15.6 ÷ 3 = ☐

4.4 ÷ 4 = ☐   10.5 ÷ 5 = ☐   12.3 ÷ 3 = ☐

0.81 ÷ 9 = ☐   42.6 ÷ 6 = ☐   0.96 ÷ 6 = ☐

0.52 ÷ 2 = ☐   0.96 ÷ 8 = ☐   40.5 ÷ 5 = ☐

20.4 ÷ 6 = ☐   11.7 ÷ 9 = ☐   25.6 ÷ 8 = ☐

STEP 3 (1.5 min) **Challenge**

Start the timer

Answer these.

12.1 ÷ 11 = ☐   13.2 ÷ 11 = ☐   14.3 ÷ 11 = ☐   15.4 ÷ 11 = ☐

1.76 ÷ 11 = ☐   14.4 ÷ 12 = ☐   15.6 ÷ 12 = ☐   16.5 ÷ 15 = ☐

Time spent: _____ min _____ sec. Total: _____ out of 34

Date: _____

Day of Week: _____

STEP 1 (1 min) Warm-up

Start the timer

Answer these.

$2.1 \div 3 =$

$3.6 \div 9 =$

$0.48 \div 6 =$

$54.9 \div 9 =$

$0.9 \div 9 =$

$0.48 \div 2 =$

$9.2 \div 4 =$

$7.5 \div 5 =$

$0.24 \div 8 =$

STEP 2 (2.5 min) Rapid calculation

Start the timer

(TIP) *To divide a number by a large divisor mentally, split the divisor into two factors to simplify the calculation. For example, $36 \div 15 = 36 \div 3 \div 5 = 12 \div 5$.*

Answer these.

$5.5 \div 11 =$

$0.77 \div 11 =$

$3.3 \div 11 =$

$4.4 \div 8 =$

$1.21 \div 11 =$

$14.3 \div 13 =$

$1.54 \div 11 =$

$13.2 \div 12 =$

$7.2 \div 12 =$

$8.4 \div 14 =$

$0.84 \div 12 =$

$1.32 \div 11 =$

$1.04 \div 13 =$

$14.4 \div 8 =$

$1.44 \div 12 =$

$6.4 \div 16 =$

$27.5 \div 11 =$

$7.2 \div 24 =$

STEP 3 (1.5 min) Challenge

Start the timer

Answer these.

$3.3 \div 15 =$

$9.43 \div 23 =$

$1.05 \div 35 =$

$94.5 \div 45 =$

$32.8 \div 16 =$

$3.36 \div 16 =$

$380 \div 76 =$

$5.6 \div 28 =$

Time spent: _____ min _____ sec. Total: _____ out of 35

Date: _____

Day of Week: _____

STEP 1 — 1 min — Warm-up

Start the timer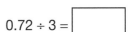

Answer these.

2.1 ÷ 30 = ☐   3.6 ÷ 9 = ☐   0.48 ÷ 6 = ☐

54.9 ÷ 9 = ☐   0.9 ÷ 9 = ☐   0.48 ÷ 12 = ☐

9.2 ÷ 8 = ☐   7.5 ÷ 15 = ☐   5.64 ÷ 8 = ☐

STEP 2 — 2.5 min — Rapid calculation

Start the timer

Answer these.

0.6 ÷ 3 = ☐   0.7 ÷ 14 = ☐   0.72 ÷ 3 = ☐

0.32 ÷ 4 = ☐   0.84 ÷ 6 = ☐   0.74 ÷ 2 = ☐

0.65 ÷ 13 = ☐   1.7 ÷ 2 = ☐   0.36 ÷ 12 = ☐

0.54 ÷ 3 = ☐   0.72 ÷ 24 = ☐   0.144 ÷ 6 = ☐

10.4 ÷ 13 = ☐   1.44 ÷ 16 = ☐   0.45 ÷ 9 = ☐

0.032 ÷ 16 = ☐   0.248 ÷ 4 = ☐   4.84 ÷ 4 = ☐

0.256 ÷ 4 = ☐   0.072 ÷ 36 = ☐

STEP 3 — 1.5 min — Challenge

Start the timer

Answer these.

4.209 ÷ 3 = ☐   0.132 ÷ 11 = ☐   0.143 ÷ 13 = ☐

0.187 ÷ 11 = ☐   1.65 ÷ 15 = ☐   0.276 ÷ 23 = ☐

3.45 ÷ 23 = ☐   0.6 ÷ 12 = ☐   0.132 ÷ 44 = ☐

Time spent: _____ min _____ sec. Total: _____ out of 38

Date: _____

Day of Week: _____

STEP 1 $\binom{1}{min}$ Warm-up

Start the timer

Answer these.

$4 \times 27 =$ ☐

$30 \div 4 =$ ☐

$3.6 \div 3 =$ ☐

$6.3 \div 7 =$ ☐

$0.96 \div 2 =$ ☐

$2.4 \div 8 =$ ☐

$12.5 \div 5 =$ ☐

$3.2 \div 8 =$ ☐

STEP 2 $\binom{2.5}{min}$ Rapid calculation

Start the timer

TIP To solve decimal multiplication and division problems involving 0.25, simplify the calculation by multiplying by 4 first, then multiply or divide the other number in the calculation by 4 as determined by the operation.

For example:

$5.6 \times 0.25 = (5.6 \div 4) \times (0.25 \times 4) = 5.6 \div 4 = 1.4$

$7.5 \div 0.25 = (7.5 \times 4) \div (0.25 \times 4) = (7.5 \times 4) \div 1 = 7.5 \times 4 = 30$

You can use the same technique for calculations involving 2.5 and 25.

Answer these.

$1.6 \times 0.25 =$ ☐

$2.4 \times 0.25 =$ ☐

$0.6 \div 0.25 =$ ☐

$92 \times 0.25 =$ ☐

$0.25 \times 48 =$ ☐

$3.6 \div 0.25 =$ ☐

$0.25 \times 3.2 =$ ☐

$4.4 \times 0.25 =$ ☐

$0.25 \times 0.52 =$ ☐

$56 \times 0.25 =$ ☐

$0.25 \times 2.8 =$ ☐

$8.8 \div 0.25 =$ ☐

$0.25 \times 26 =$ ☐

$1.8 \times 0.25 =$ ☐

$6.4 \div 0.25 =$ ☐

$30 \times 0.25 =$ ☐

$0.25 \times 16 =$ ☐

$4 \div 0.25 =$ ☐

STEP 3 $\binom{1.5}{min}$ Challenge

Start the timer

Answer these.

$7.2 \times 0.25 =$ ☐

$6.4 \times 0.25 =$ ☐

$18 \div 0.25 =$ ☐

$52 \times 2.5 =$ ☐

$2.5 \times 4.8 =$ ☐

$50 \div 2.5 =$ ☐

$2.5 \times 1.8 =$ ☐

$6 \div 2.5 =$ ☐

Time spent: _____ min _____ sec. Total: _____ out of 34

Date: _____

Day of Week: _____

STEP 1 (1 min) Warm-up

Answer these.

2.4 ÷ 0.8 = ☐   0.36 ÷ 0.4 = ☐   0.64 ÷ 0.8 = ☐   0.72 ÷ 3.6 = ☐

10 ÷ 20 = ☐   12.8 ÷ 4 = ☐   12.5 ÷ 5 = ☐   4.8 ÷ 0.3 = ☐

STEP 2 (2.5 min) Rapid calculation

Start the timer

(TIP) *If there are no brackets in a multiplication, division or mixed multiplication/division sentence containing three or more numbers, changing the sequence of calculation will not change the result.*

For example: 18 × 24 ÷ 9 = 18 ÷ 9 × 24 = 48

Dividing a dividend by two numbers in succession gives the same result as dividing it by the product of the two divisors.

For example: 52 ÷ 25 ÷ 4 = 52 ÷ (25 × 4) = 52 ÷ 100 = 0.52

Answer these.

14 ÷ 0.25 ÷ 10 = ☐   19 ÷ 0.25 ÷ 4 = ☐   27 ÷ 0.125 ÷ 8 = ☐

4.3 ÷ 5 ÷ 2 = ☐   24 ÷ 0.2 ÷ 5 = ☐   14 ÷ 0.5 ÷ 2 = ☐

6 ÷ 2.5 ÷ 4 = ☐   23 ÷ 2.5 ÷ 0.4 = ☐   5.2 ÷ 1.3 ÷ 4 = ☐

5.4 ÷ 0.3 ÷ 0.9 = ☐   24 ÷ 5 ÷ 6 = ☐   16 ÷ 0.8 ÷ 4 = ☐

56 ÷ 16 ÷ 7 = ☐   32 ÷ 0.8 ÷ 0.4 = ☐   3.6 ÷ 0.9 ÷ 0.2 = ☐

STEP 3 (1.5 min) Challenge

Start the timer

Answer these.

42 ÷ 0.2 ÷ 6 = ☐   0.64 ÷ 0.8 ÷ 4 = ☐   0.84 ÷ 6 ÷ 0.7 = ☐

12.6 ÷ 14 ÷ 3 = ☐   0.72 ÷ 0.2 ÷ 0.9 = ☐   3.6 ÷ 6 ÷ 2 = ☐

54 ÷ 3 ÷ 0.9 = ☐   18 × 25 ÷ 9 = ☐

Time spent: _____ min _____ sec. Total: _____ out of 31

Date: _____

Day of Week: _____

STEP 1 (1 min) **Warm-up**

Start the timer

Answer these.

0.8 + 0.4 + 0.2 = ☐

1.3 + 0.9 + 0.7 = ☐

2.6 + 1.2 + 0.4 = ☐

1.9 + 0.5 + 3.5 = ☐

9.4 + 7 + 3 = ☐

2.5 × 7 × 0.4 = ☐

4 × 21 × 0.5 = ☐

8 × 3 × 0.125 = ☐

0.5 × 6 + 1.5 × 6 = ☐

STEP 2 (2.5 min) **Rapid calculation**

Start the timer

Answer these.

12.5 + 3.4 + 7.5 = ☐

3.8 + 2.1 + 4.9 = ☐

5.3 + 1.8 + 2.7 = ☐

6.3 + 9.4 + 4.6 = ☐

5.4 + 4.6 + 6.7 = ☐

2.5 + 1.7 + 7.5 = ☐

1.9 + 2.1 + 3.5 = ☐

2.7 + 0.9 + 8.3 = ☐

2.5 × 0.2 × 4 = ☐

1.25 × 8 × 1.2 = ☐

4 × 1.2 × 2.5 = ☐

0.04 × 2.5 + 12.5 = ☐

(12.5 − 5) × 0.8 = ☐

6.5 × 3.9 + 6.1 × 6.5 = ☐

7.4 × 99 + 7.4 = ☐

10 × 8 − 0.8 = ☐

STEP 3 (1.5 min) **Challenge**

Start the timer

Answer these.

4.6 − 2.7 + 1.4 = ☐

8 × 0.6 ÷ 0.4 = ☐

1.5 × 28 ÷ 0.7 = ☐

8.8 × 2.5 = ☐

1.01 × 46 = ☐

11 × 4.5 = ☐

(10 − 0.1) × 3.8 = ☐

3.2 × 1.25 = ☐

Time spent: _____ min _____ sec. Total: _____ out of 33

Date: _____

Day of Week: _____

STEP 1 (1 min) Warm-up

Start the timer

Convert the decimals into fractions. Use mixed numbers where appropriate.

0.2 = ☐   0.951 = ☐   0.27 = ☐   0.347 = ☐

8.6 = ☐   0.43 = ☐   0.06 = ☐   0.008 = ☐

2.36 = ☐   5.021 = ☐

STEP 2 (2.5 min) Rapid calculation

Start the timer

Convert these into decimals.

$\frac{3}{10}$ = ☐ $\frac{48}{100}$ = ☐ $\frac{529}{1000}$ = ☐

$\frac{75}{1000}$ = ☐ $\frac{7}{100}$ = ☐ $\frac{21}{10}$ = ☐

$\frac{392}{100}$ = ☐ $2\frac{408}{1000}$ = ☐ $7\frac{2}{1000}$ = ☐

$1\frac{64}{1000}$ = ☐ $\frac{2}{5}$ = ☐ $\frac{3}{8}$ = ☐

$\frac{7}{50}$ = ☐ $\frac{4}{25}$ = ☐ $1\frac{9}{20}$ = ☐

STEP 3 (1.5 min) Challenge

Start the timer

Convert these into decimals.

$\frac{11}{25}$ = ☐ $\frac{5}{8}$ = ☐ $\frac{80}{32}$ = ☐

$\frac{28}{50}$ = ☐ $\frac{6}{25}$ = ☐ $2\frac{1}{4}$ = ☐

$\frac{23}{5}$ = ☐ $\frac{63}{28}$ = ☐

Time spent: _____ min _____ sec. Total: _____ out of 33

STEP 1 (1 min) Warm-up

Start the timer

Answer these using fractions, mixed numbers or decimals.

$\frac{2}{3} + 0.5 =$ ☐

$2\frac{3}{4} + 1.2 =$ ☐

$\frac{3}{4} \div 3 =$ ☐

$\frac{3}{4} \times \frac{2}{9} =$ ☐

$3\frac{1}{8} - 1.2 =$ ☐

$3\frac{1}{2} - 1.7 =$ ☐

$\frac{5}{8} \div 10 =$ ☐

$\frac{3}{16} \times \frac{4}{15} =$ ☐

STEP 2 (2.5 min) Rapid calculation

Start the timer

Answer these using fractions, mixed numbers or decimals.

$1\frac{3}{5} - 0.7 =$ ☐

$3\frac{1}{2} - 0.375 =$ ☐

$5\frac{1}{8} + 3.5 =$ ☐

$\frac{9}{26} \times \frac{13}{30} =$ ☐

$\frac{28}{45} \times \frac{15}{56} =$ ☐

$8\frac{1}{2} \times 0.7 =$ ☐

$\frac{5}{24} \div 25 =$ ☐

$5\frac{4}{25} + 4.25 =$ ☐

$\frac{7}{10} \times 1.8 =$ ☐

$\frac{5}{14} \times 2 =$ ☐

$8\frac{7}{12} - 4\frac{1}{3} =$ ☐

$\frac{15}{38} \div 20 =$ ☐

STEP 3 (1.5 min) Challenge

Start the timer

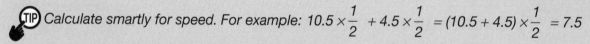 **TIP** Calculate smartly for speed. For example: $10.5 \times \frac{1}{2} + 4.5 \times \frac{1}{2} = (10.5 + 4.5) \times \frac{1}{2} = 7.5$

Answer these using fractions, mixed numbers or decimals.

$3\frac{3}{4} \times 6.5 + 2\frac{1}{4} \times 6.5 =$ ☐

$6.1 \times \frac{3}{4} - 3.1 \times \frac{3}{4} =$ ☐

$\frac{5}{12} \times \frac{3}{7} + \frac{5}{12} \times \frac{4}{7} =$ ☐

$7.5 \times 5\frac{1}{2} + 7\frac{1}{2} \times 3.5 + 7\frac{1}{2} =$ ☐

$16.25 \times \frac{11}{12} - 13\frac{1}{4} \times \frac{11}{12} =$ ☐

$13\frac{7}{8} \times \frac{5}{22} - 3.875 \times \frac{5}{22} =$ ☐

$15.2 \times \frac{1}{6} + 4\frac{4}{5} \times \frac{1}{6} =$ ☐

$3\frac{7}{10} \times 1.5 + 8\frac{1}{2} \times 3.7 =$ ☐

Time spent: _____ min _____ sec. Total: _____ out of 28

©HarperCollins*Publishers* 2019

Date: _____

Day of Week: _____

STEP 1 (1 min) Warm-up

Start the timer

Answer these using fractions, mixed numbers or decimals.

$2\frac{2}{5} + 2.5 =$ ☐

$9\frac{3}{4} + 1.8 =$ ☐

$\frac{25}{36} \div 15 =$ ☐

$\frac{14}{45} \times \frac{15}{16} =$ ☐

$12\frac{5}{8} - 4.4 =$ ☐

$6\frac{1}{2} - 2.35 =$ ☐

$\frac{35}{56} \div 28 =$ ☐

$\frac{50}{81} \times \frac{72}{75} =$ ☐

STEP 2 (2.5 min) Rapid calculation

Start the timer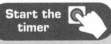

Answer these using fractions, mixed numbers or decimals.

$3\frac{1}{5} \times 0.25 =$ ☐

$5\frac{1}{4} \times 0.8 =$ ☐

$\frac{45}{94} \div 25 =$ ☐

$5\frac{7}{25} + 4.5 =$ ☐

$\frac{42}{45} \times \frac{15}{14} \div 9 =$ ☐

$\frac{5}{42} \times 28 =$ ☐

$10\frac{5}{6} - 6\frac{5}{12} =$ ☐

$\frac{35}{46} \div 20 =$ ☐

$6\frac{3}{5} - 3.8 + \frac{2}{5} =$ ☐

$3\frac{1}{2} - 1.7 - 1.3 =$ ☐

$6\frac{5}{8} + 3.5 - 1\frac{3}{8} =$ ☐

$\frac{15}{52} \times \frac{39}{150} \div 2 =$ ☐

STEP 3 (1.5 min) Challenge

Start the timer

Answer these using fractions, mixed numbers or decimals.

$2\frac{3}{8} \times 16.4 + 7\frac{5}{8} \times 16.4 =$ ☐

$32.5 \times \frac{3}{4} - 12.5 \times \frac{3}{4} =$ ☐

$\frac{7}{32} \times 1\frac{3}{17} + \frac{7}{32} \times \frac{14}{17} =$ ☐

$7.08 \times 5\frac{1}{2} + 7\frac{2}{25} \times 3.5 + 7\frac{2}{25} =$ ☐

$63.75 \times \frac{5}{12} - 3\frac{3}{4} \times \frac{5}{12} =$ ☐

$26\frac{5}{8} \times \frac{13}{22} - 4.625 \times \frac{13}{22} =$ ☐

$19.4 \times \frac{3}{8} + 4\frac{3}{5} \times 0.375 =$ ☐

$3\frac{1}{5} \times 24.5 - 14\frac{1}{2} \times 3.2 =$ ☐

Date: _____

Day of Week: _____

STEP 1 (1 min) Warm-up

Start the timer

Simplify these expressions. The first two have been done for you.

$a + a =$ **2a** $\qquad$ $9 \times z =$ **9z** $\qquad$ $b + b =$ ☐ $\qquad$ $4 \times d =$ ☐

$c + c + c =$ ☐ $\qquad$ $3y - y =$ ☐ $\qquad$ $5r - 2r =$ ☐ $\qquad$ $f \times 5 =$ ☐

$7s - 3s =$ ☐ $\qquad$ $t + t + t + t + t =$ ☐

STEP 2 (2.5 min) Rapid calculation

Start the timer

Simplify these expressions.

$1 \times a =$ ☐ $\qquad$ $8 \times b =$ ☐ $\qquad$ $2z + 5z + 3z =$ ☐ $\qquad$ $9p - 3p =$ ☐

$g \times 12 =$ ☐ $\qquad$ $4v - 3v =$ ☐ $\qquad$ $0.5r + 0.5r =$ ☐ $\qquad$ $5n + 7n - 3n =$ ☐

$15 \times e =$ ☐ $\qquad$ $6y \times 2 =$ ☐ $\qquad$ $3x \div x =$ ☐ $\qquad$ $12t \div 2 =$ ☐

$b + 2b =$ ☐ $\qquad$ $54a \div 2 =$ ☐ $\qquad$ $3.5b + b =$ ☐ $\qquad$ $12x \div 12 =$ ☐

STEP 3 (1.5 min) Challenge

Start the timer

Simplify these expressions.

$19m - 4m =$ ☐ $\qquad$ $7s \times 8 =$ ☐ $\qquad$ $5a + 7a =$ ☐ $\qquad$ $3 \times 2.4p =$ ☐

$9.5t - 2t =$ ☐ $\qquad$ $7.2a - 0.2a =$ ☐ $\qquad$ $9 \times 7s =$ ☐ $\qquad$ $43t - t =$ ☐

$24x + 9x =$ ☐ $\qquad$ $7.6y - 6y =$ ☐ $\qquad$ $10.3t - 3.8t =$ ☐ $\qquad$ $36s \div 3 =$ ☐

$5 \times 1.2x =$ ☐ $\qquad$ $13 \times 7m =$ ☐ $\qquad$ $35k - 20k =$ ☐ $\qquad$ $17.6y - 6.6y =$ ☐

Time spent: _____ min _____ sec. Total: _____ out of 40

STEP 1 (1 min) Warm-up

Start the timer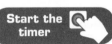

Simplify these.

$9b + 4b =$ ☐ $21a - 7a =$ ☐ $60h - 45h =$ ☐ $9 \times a =$ ☐

$y \times y =$ ☐ $12 \times 9c =$ ☐ $15 \times c =$ ☐ $b \times 1 =$ ☐

$3x + 4.5x =$ ☐ $3.2x - x =$ ☐

STEP 2 (2.5 min) Rapid calculation

Start the timer

Simplify these.

$42k + 18k + 9k =$ ☐ $100g - 42g - 58g =$ ☐ $80m - 36m + 10m =$ ☐

$24d + 12d - 5d =$ ☐ $6s \times 2 \times 4 =$ ☐ $5 \times 4t \times 6 =$ ☐

$8.09a - 3.53a =$ ☐ $42y + 16y - 24y =$ ☐ $5p \times 1.5 =$ ☐

$18q \times 0.5 =$ ☐ $25x \div 5 \times 12 =$ ☐ $50c \times 5 \div 25 =$ ☐

STEP 3 (1.5 min) Challenge

Start the timer

Simplify these.

$4 \times 8q - 30q =$ ☐ $46b \div 2 \div 23 =$ ☐ $2y \times 18 \div 4 =$ ☐

$12x + 1.2x + 2.8x =$ ☐ $54a \div 6 + a =$ ☐ $49r - 49r \div 7 =$ ☐

$3 \times 4a + 7a =$ ☐ $3 \times 2.4p - 1.2p =$ ☐

33 Simple equations (1)

Date: _____

Day of Week: _____

STEP 1 (1 min) Warm-up

Start the timer

Solve these equations.

$6 + x = 13$

$x = $ ☐

$15 - x = 7$

$x = $ ☐

$x + 18 = 36$

$x = $ ☐

$x - 4 = 11$

$x = $ ☐

$29 + x = 40$

$x = $ ☐

$2.7 - x = 0.8$

$x = $ ☐

$40.5 - x = 10$

$x = $ ☐

$x - 5.7 = 4$

$x = $ ☐

STEP 2 (2.5 min) Rapid calculation

Start the timer

Solve these equations.

$x - 5.7 = 12.5$

$x = $ ☐

$66 + x = 100$

$x = $ ☐

$x + 22.5 = 40$

$x = $ ☐

$31.4 - x = 21$

$x = $ ☐

$x - 91 = 109$

$x = $ ☐

$x + 21 = 31$

$x = $ ☐

$4.6 + x = 6.4$

$x = $ ☐

$x - 0.9 = 3.6$

$x = $ ☐

$66.7 - x = 30.7$

$x = $ ☐

$75 - x = 75$

$x = $ ☐

$x - 30 = 40$

$x = $ ☐

$x - 0.9 = 10.9$

$x = $ ☐

$x + 8.5 = 19.3$

$x = $ ☐

$22 + x = 43.4$

$x = $ ☐

$x + 13.1 = 29$

$x = $ ☐

STEP 3 (1.5 min) Challenge

Start the timer

Solve these equations.

$12 + 4 + x = 20$

$x = $ ☐

$x + 6.7 = 10 - 2$

$x = $ ☐

$x + 2.4 = 6 - 0.5$

$x = $ ☐

$x - (10 + 0.4) = 21$

$x = $ ☐

$x - 17 = 70 - 17$

$x = $ ☐

$x + 0.8 = 0.3 \times 3$

$x = $ ☐

Time spent: _____ min _____ sec. Total: _____ out of 29

STEP 1 (1 min) **Warm-up**

Start the timer

Solve these.

$3x = 18$

$x = \boxed{}$

$10 + x = 36$

$x = \boxed{}$

$70 - x = 51$

$x = \boxed{}$

$63 \div x = 3$

$x = \boxed{}$

$x \div 11 = 3$

$x = \boxed{}$

$2x + 8 = 16$

$x = \boxed{}$

$6x - 8 = 4$

$x = \boxed{}$

$2x - 6 = 12$

$x = \boxed{}$

STEP 2 (2.5 min) **Rapid calculation**

Start the timer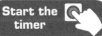

Solve these.

$6x + 2 = 20$

$x = \boxed{}$

$10 \div x + 4 = 9$

$x = \boxed{}$

$30 - 4x = 10$

$x = \boxed{}$

$72 - 2x = 52$

$x = \boxed{}$

$150 \div x = 25$

$x = \boxed{}$

$x \div 0.5 = 2.6$

$x = \boxed{}$

$12 + 34x = 80$

$x = \boxed{}$

$2 - 6x = 0.2$

$x = \boxed{}$

$6x - 6.3 = 5.7$

$x = \boxed{}$

$19 - 14x = 12$

$x = \boxed{}$

$x \div 1.5 = 4$

$x = \boxed{}$

$6x \div 4 = 15$

$x = \boxed{}$

$4x + 3.4 = 5.4$

$x = \boxed{}$

$11x + 25 - 14 = 99$

$x = \boxed{}$

$15x \div 10 = 7.5$

$x = \boxed{}$

STEP 3 (1.5 min) **Challenge**

Start the timer

Solve these.

$4 \times 25 - x = 4$

$x = \boxed{}$

$5x + 24 \div 4 = 126$

$x = \boxed{}$

$20 = 6x - 34$

$x = \boxed{}$

$8 \times 9 - 2x = 10$

$x = \boxed{}$

$42 - 6 = 4x + 4$

$x = \boxed{}$

$9.6 \div x = 4 + 2$

$x = \boxed{}$

©HarperCollins*Publishers* 2019

Time spent: _____ min _____ sec. Total: _____ out of 29

37

35 Ratio

Date: _____

Day of Week: _____

STEP 1 (1 min) Warm-up
Start the timer

Simplify the ratios.

$15:3 =$ ⬜ $14:35 =$ ⬜ $42:54 =$ ⬜ $54:63 =$ ⬜

$8:12 =$ ⬜ $9:15 =$ ⬜ $39:45 =$ ⬜ $92:161 =$ ⬜

STEP 2 (2.5 min) Rapid calculation
Start the timer

Simplify the ratios.

$4.2:2.4 =$ ⬜ $3.25:0.15 =$ ⬜ $0.5:2.5 =$ ⬜ $0.9:0.72 =$ ⬜

$\dfrac{2}{9}:\dfrac{1}{3} =$ ⬜ $0.4:\dfrac{3}{4} =$ ⬜ $\dfrac{3}{4}:\dfrac{1}{6} =$ ⬜ $1\dfrac{1}{2}:\dfrac{1}{3} =$ ⬜

$1\dfrac{3}{5}:0.8 =$ ⬜ $\dfrac{1}{2}:\dfrac{3}{4} =$ ⬜ $\dfrac{7}{8}:35 =$ ⬜ $36:\dfrac{4}{5} =$ ⬜

STEP 3 (1.5 min) Challenge
Start the timer

Simplify the ratios.

$30\,min:1.5\,h =$ ⬜ $4\,kg:500\,g =$ ⬜ $10\,days:2\,weeks =$ ⬜

$3\,m:40\,cm =$ ⬜ $800\,g:2\,kg =$ ⬜ $25\,cm:5\,m =$ ⬜

$15\,h:1\,day =$ ⬜ $1\,min:45\,sec =$ ⬜

🧠 Mind Gym

Sudoku is a nine-block (3 × 3) puzzle in which each block is subdivided into nine cells. Complete the puzzle using each of the digits from 1 to 9 once only in each block so that each row and column also contains each of the digits from 1 to 9 once only.

		6		7		5		
	7			8	1			6
			5			1		4
6					8		9	
				5				
	9		3					1
8		2			7			
5			9	3			6	
		9		1		4		

Time spent: _____ min _____ sec. Total: _____ out of 28

©HarperCollins*Publishers* 2019

STEP 1 (1 min) Warm-up

Start the timer

Convert these to percentages.

$\frac{1}{2}$ = [ ] $\frac{1}{8}$ = [ ] $\frac{1}{16}$ = [ ] $\frac{1}{20}$ = [ ]

$\frac{1}{25}$ = [ ] $\frac{1}{4}$ = [ ] $\frac{3}{10}$ = [ ] $\frac{1}{5}$ = [ ]

STEP 2 (2.5 min) Rapid calculation

Start the timer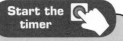

 TIP *To convert a decimal to a percentage, move the digits two place values to the left and write the percentage symbol behind it. For example, 0.52 becomes 52%.*

Convert these to percentages.

$\frac{2}{5}$ = [ ] $\frac{3}{8}$ = [ ] $\frac{5}{16}$ = [ ] 0.45 = [ ]

3.7 = [ ] $\frac{4}{25}$ = [ ] $2\frac{1}{4}$ = [ ] $1\frac{3}{5}$ = [ ]

$\frac{7}{20}$ = [ ] 0.087 = [ ] $7\frac{4}{5}$ = [ ] 0.12 = [ ]

$5\frac{2}{5}$ = [ ] 1.0 = [ ] 0.2 = [ ] 0.01 = [ ]

STEP 3 (1.5 min) Challenge

Start the timer

Fill in the missing percentages.

1. The ratio of apples to oranges is 4:1. The percentage of oranges is [ ] %.

2. A quiz contained exactly 30 questions. Jack answered six questions incorrectly.

Jack's accuracy was [ ] %.

3. Ryan weighs 40 kg and Harry weighs 55 kg. Harry is [ ] % heavier than Ryan.

4. If a fruit shop sells apples at 65% of the original price, the current price of the apples is [ ] % lower than the original price.

Time spent: _____ min _____ sec. Total: _____ out of 28

Date: _____

Day of Week: _____

Tick the true statement(s).

1. The teacher announced that the average score for an exam was 82.

   This means that all the students in the class got 82. ☐

2. There are five stacks of exercise books. They have an average of 40 books per stack.

   One of the stacks could have 20 copies. ☐

3. The average depth of a pond is 140 cm.

   Lucy is 155 cm tall so she will definitely be able to walk across the pond. ☐

Calculate the mean of each set of numbers.

26, 30, 27, 29 ☐   75, 78, 65, 74 ☐   18, 18, 24, 25, 20, 21 ☐

15, 10, 12, 16, 14 ☐   29, 25, 23, 21, 20, 26 ☐   24, 28, 20, 17, 23 ☐

Fill in the missing numbers.

1. The table shows the number of stars that each of five children cut out.

Name	Rubina	Albert	Alice	Kitty	Tom
Number of stars cut out	78	82	65	90	85

The children cut out an average of ☐ stars each.

2. The table shows the amount of water that Jamie's family consumed each month for six months.

Month	1	2	3	4	5	6
Amount of water used (m³)	84	90	84	85	86	90

Their average monthly consumption of water is ☐ m³.

3. The table shows the number of students in seven groups.

Group	1	2	3	4	5	6	7
Number of students	42	48	43	40	42	49	44

There is an average of ☐ students per group.

Time spent: _____ min _____ sec. Total: _____ out of 12

STEP 1 (1 min) Warm-up

Start the timer

Choose the correct options from the box to complete the sentences.

perpendicular to each other	**perpendicular line**	**parallel lines**
perpendicular foot	**parallel to each other**	

Two lines in a plane that do not meet are called ....,

which means the two lines are ....

If two lines meet at a right angle, the two lines are ....

One of the lines is the .... of the other and the

intersection point of the lines is called the ....

STEP 2 (2.5 min) Rapid calculation

Start the timer

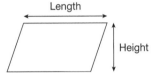
Length

Height

Use the information given about each parallelogram to work out the unknown value.

Length (cm)	77	18	35	46	12	9.1	2.2	4.6	5.5	1.3
Height (cm)	2	8	6	5	13	9	6	3	3	7
Area (cm²)										

Length (m)	14							3	4	13
Height (m)	11	15	2	40	5	4	9			
Area (m²)		45	4	80	2.5	2.4	2.7	12	14	52

STEP 3 (1.5 min) Challenge

Start the timer

Use the information given about each parallelogram to work out the unknown value.

Length (m)	0.25		15		60		30		
Height (m)		0.2		12.5		2		36	56
Area (m²)	5	4	6	200	420	3.8	12	14.4	16.8

Date: _____

Day of Week: _____

STEP 1 (1 min) **Warm-up** — Start the timer

 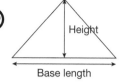

Height

Base length

Use the information given about each triangle to work out its area.

Base length (cm)	12	10	25	32	45	34	75	100	10
Height (cm)	20	102	2	4	6	5	20	25	5
Area (cm²)									

STEP 2 (2.5 min) **Rapid calculation** — Start the timer

Use the information given about each triangle to work out its area.

Base length (cm)	24	12	20	20	5	12	20	80
Height (cm)	10	11	13	25	3	4	15	25
Area (cm²)								

Base length (m)	54	5.4	1.2	6.8	5.8	9.3	4.8
Height (m)	30	6	5	6	5	4	4
Area (m²)							

STEP 3 (1.5 min) **Challenge** — Start the timer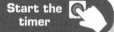

Use the information given about each triangle to work out its area.

Base length (cm)	16	25	3.1	4.3	5.6
Height (cm)	9	4.4	8	10	7
Area (cm²)					

Base length (m)	66	7.8	8.8	7.8	
Height (m)	4	8	5	5	6
Area (m²)					75

Time spent: _____ min _____ sec. Total: _____ out of 34

Date: _____

Day of Week: _____

STEP 1 (1 min) Warm-up

Start the timer

Answer these using fractions or mixed numbers.

$\frac{8}{33} \div 4 =$ ☐

$\frac{7}{12} \times \frac{2}{9} =$ ☐

$5 - 2\frac{3}{4} =$ ☐

$\frac{1}{5} + \frac{1}{6} =$ ☐

$\frac{19}{14} \div 38 =$ ☐

$7\frac{3}{4} - 4\frac{3}{11} =$ ☐

$8\frac{2}{9} + 12\frac{7}{18} =$ ☐

$\frac{27}{56} \times \frac{14}{45} =$ ☐

STEP 2 (2.5 min) Rapid calculation

Start the timer

Tick the nets that will form a cube.

1.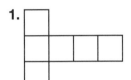

2.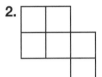

3.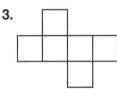

4.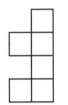

5.

6.

7.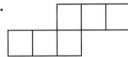

8.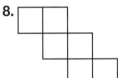

9.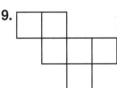

STEP 3 (1.5 min) Challenge

Start the timer

These nets all form a cube when folded. Number each face so that the faces that will be opposite each other have the same number.

1.

2.

3.

4.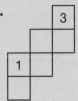

5.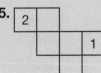

6.

Time spent: _____ min _____ sec. Total: _____ out of 23

ANSWERS

Answers are given from top left, left to right, unless otherwise stated.

Test 1

Step 1:

From top:

3 000 000 + 500 000 + 70 000 + 1000 + 500 + 90 + 6;

9 000 000 + 400 000 + 30 000 + 6000 + 0 + 30 + 4;

6 000 000 + 900 000 + 90 000 + 3000 + 900 + 0 + 5;

2 000 000 + 200 000 + 80 000 + 5000 + 400 + 30 + 9;

Step 2:

Table completed as follows from top: 3 423 704;

7 293 868; 9 379 020; 5 028 309; 4 603 108; 1 007 004

Step 3:

First row: 9 378 000; nine million, three hundred and seventy-eight thousand

Second row: 8 520 019; eight million, five hundred and twenty thousand and nineteen

Third row: 6 807 390; six million, eight hundred and seven thousand, three hundred and ninety

Test 2

Step 1:

4 480 000; 1 740 000; 5 010 000; 7 400 000; 9 530 000; 8 240 000; 2 440 000; 3 550 000

Step 2:

3 000 000; 6 600 000; 5 500 000; 4 900 000; 1 700 000; 7 500 000; 8 000 000; 3 800 000; 5 400 000; 6 600 000; 9 500 000; 1 200 000

Step 3:

First row: 4 500 000; 4 500 000; 4 000 000

Second row: 6 370 000; 6 400 000; 6 000 000

Third row: 2 350 000; 2 300 000; 2 000 000

Fourth row: 7 840 000; 7 800 000; 8 000 000

Test 3

Step 1:

From left: 7690; 21 380; 12 360; 22 386

Step 2:

6886; 14 833; 13 648; 13 134; 9372; 26 272; 21 906; 22 685

Step 3:

From left: 145 204; 255 584; 222 888; 463 344

Test 4

Step 1:

95; 88; 85; 96; 59; 74

Step 2:

3524; 437; 1140; 493; 1411; 2413; 428; 1170; 856

Step 3:

237 r2; 1541 r2; 635 r5; 648 r3; 760 r2; 477 r4; 1388 r4; 1060 r0

Test 5

Step 1:

26; 17; 19; 25; 30; 34

Step 2:

239; 49; 587; 65; 150; 301; 589; 383; 245

Step 3:

74 r23; 19 r4; 112 r14; 132 r13; 68 r54; 123 r45

Test 6

Step 1:

664; 495; 613; 90; 212; 174; 425; 170

Step 2:

664; 491; 174; 4950; 650; 84; 78; 685; 7600; 19; 96; 170; 2250; 2820; 850

Step 3:

60; 2400; 234; 10; 491; 431; 90; 212; 85

Test 7

Step 1:

1.   From left to right: 8; 20; 68

2.   16; 16; 44

Step 2:

1.   1.2; 2.4; 5.6; 18; 11.6; 0.56; 64; 44.8

2.   6.8; 4; 8; 16.4

Step 3:

1.   From left to right: 0.8; 1.1; 0.04

2.   0.8; 1; 3.9

Test 8

Step 1:

48; 96; 25; 6; 32; 6

Step 2:

4.2; 1.8; 0.09; 0.6; 1.6; 2.6; 15; 16.8; 6.8; 10

Step 3:

0.01; 1350; 12 100; 4; 2000; 1.7

Test 9

Step 1:

$\frac{5}{13}$; 5; $\frac{4}{7}$; 11, 6; 7, 3; $\frac{6}{17}$; 2; $\frac{7}{15}$

Step 2:

$\frac{3}{11}$; 8; $\frac{14}{17}$; 7, 12; 36, 13; $\frac{2}{5}$; 2; $\frac{27}{25}$; $\frac{15}{7}$; 15; $\frac{17}{6}$; 9, 2; 4, 7; $\frac{4}{18}$;

8; $\frac{3}{11}$; 18; 12, 5; 15; 20, 3

Step 3:

Equivalent answers are possible.

$\frac{37}{60}$; $\frac{19}{1000}$; $\frac{3}{100}$; $\frac{27}{24}$; $\frac{53}{1000}$; $\frac{327}{1000}$; $\frac{17}{100}$; $\frac{50}{1000}$

Test 10

Step 1:

4; 2; 20; 6; 10; 4; 32; 5

Step 2:

6; 2; 24; 6; 5; 8; 12; 9; 36; 4; 15; 35; 8; 42; 4

Step 3:

2; 2; 10; 7; 60; 75; 20; 5

Test 11

Step 1:

$\frac{8}{24} = \frac{16}{48} = \frac{1}{3} = 1 \div 3$; $\frac{15}{120} = \frac{6}{48} = 0.125 = 1 \div 8$;

$\frac{24}{108} = \frac{18}{81} = \frac{2}{9} = 12 \div 54$; $\frac{20}{25} = \frac{32}{40} = 0.8 = 24 \div 30$

Step 2:

$\frac{1}{2}$; $\frac{2}{3}$; $\frac{1}{4}$; $\frac{1}{4}$; $\frac{2}{5}$; $\frac{4}{9}$; $\frac{4}{9}$; $\frac{2}{11}$; $\frac{3}{5}$; $\frac{7}{12}$; $\frac{3}{7}$; $\frac{2}{3}$

Step 3:

$\frac{4}{21}$; $\frac{2}{3}$; $\frac{8}{19}$; $\frac{7}{29}$; $\frac{2}{13}$; $\frac{3}{4}$; $\frac{5}{11}$; $\frac{3}{25}$

Test 12

Step 1:

=; <; >; >; =; >; >; <

Step 2:

>; >; <; >; <; >; <; >; >; >; >; <; >; >; >; >

Step 3:

1. $\frac{2}{5}$; $\frac{1}{6}$; $\frac{3}{8}$; $\frac{4}{9}$; $\frac{3}{7}$; $\frac{5}{12}$

2. $\frac{6}{7}$; $\frac{6}{11}$; $\frac{7}{10}$; $\frac{3}{5}$; $\frac{2}{3}$; $\frac{3}{4}$

Test 13

Equivalent answers are possible.

Step 1:

$\frac{1}{2}$, $\frac{14}{9}$, 2, $\frac{5}{9}$, $\frac{1}{2}$, $\frac{14}{39}$, $\frac{5}{9}$, $\frac{3}{4}$

Step 2:

2; 2; $\frac{4}{3}$; $\frac{1}{3}$; $\frac{1}{3}$; $\frac{1}{5}$; $\frac{9}{5}$; $\frac{2}{9}$; $\frac{1}{4}$; $\frac{1}{2}$; $\frac{4}{35}$; $\frac{3}{5}$; $\frac{3}{5}$; $\frac{4}{5}$; $\frac{2}{7}$; $\frac{1}{2}$; $\frac{2}{5}$; $\frac{3}{2}$; $\frac{6}{17}$; $\frac{3}{2}$

Step 3:

$\frac{11}{18}$; $\frac{11}{12}$; $\frac{3}{8}$; $\frac{1}{2}$; 1, $\frac{7}{8}$; $\frac{13}{12}$; $\frac{1}{2}$

Test 14

Equivalent answers are possible.

Step 1:

$\frac{16}{15}$; $\frac{16}{35}$; $\frac{53}{63}$; $\frac{1}{2}$; $\frac{39}{35}$; $\frac{1}{2}$; $\frac{85}{77}$; $\frac{19}{36}$

Step 2:

$\frac{7}{8}$; $\frac{1}{4}$; $\frac{29}{21}$; $\frac{3}{8}$; $\frac{13}{21}$; $\frac{3}{8}$; $\frac{11}{10}$; $\frac{7}{18}$; $\frac{38}{45}$; $\frac{17}{24}$; $\frac{1}{4}$; $\frac{5}{9}$; $\frac{1}{2}$; $\frac{19}{12}$; $\frac{29}{72}$;

$\frac{13}{12}$; $\frac{5}{12}$; $\frac{31}{35}$; $\frac{1}{2}$; $\frac{29}{35}$

Step 3:

$\frac{11}{8}$; $\frac{1}{6}$; $\frac{6}{7}$; $\frac{1}{10}$; $\frac{23}{18}$; $\frac{1}{7}$; $\frac{17}{18}$; $\frac{1}{5}$

Test 15

Equivalent answers are possible.

Step 1:

$\frac{1}{6}$; $\frac{1}{20}$; $\frac{9}{28}$; $\frac{5}{36}$; $\frac{3}{40}$; $\frac{10}{33}$; $\frac{5}{42}$; 28; 16

Step 2:

$\frac{14}{45}$; $\frac{5}{64}$; $\frac{8}{39}$; $\frac{5}{6}$; $\frac{1}{6}$; 28; $\frac{2}{15}$; $\frac{3}{7}$; 20; $\frac{5}{12}$; 44; $\frac{1}{12}$

Step 3:

$\frac{2}{35}$; $1\frac{2}{3}$; 1; $\frac{1}{27}$; $\frac{1}{9}$; $\frac{2}{3}$; $\frac{3}{16}$; $\frac{1}{14}$

Test 16

Step 1:

7; 4; 24; 3; 5; 14; 8; 6; 19

Step 2:

$\frac{17}{18}$; $\frac{7}{15}$; $\frac{11}{19}$; $\frac{24}{37}$; $\frac{6}{11}$; $2\frac{4}{7}$; $2\frac{1}{12}$; $4\frac{3}{5}$; $4\frac{2}{9}$; $\frac{13}{28}$; $2\frac{2}{13}$; $2\frac{17}{39}$;

$2\frac{11}{32}$; $\frac{29}{35}$; $4\frac{5}{9}$

Step 3:

Equivalent answers are possible.

1. $23 \div 21$; $42 \div 95$; $5 \div 18$; $7 \div 26$

2. $2\frac{1}{9}$; $2\frac{2}{3}$; $3\frac{10}{13}$; $2\frac{2}{11}$; $13\frac{1}{3}$; $5\frac{3}{7}$

Test 17

Equivalent answers are possible.

Step 1:

$\frac{1}{9}$; $\frac{1}{18}$; $\frac{1}{9}$; $\frac{1}{28}$; $\frac{2}{11}$; $\frac{1}{4}$; $\frac{2}{9}$; $\frac{1}{6}$; $\frac{1}{10}$

Step 2:

$\frac{2}{9}$; $\frac{1}{7}$; $\frac{1}{54}$; $\frac{5}{64}$; $\frac{2}{9}$; $\frac{3}{56}$; $\frac{2}{13}$; $\frac{1}{9}$; $\frac{3}{28}$; $\frac{1}{23}$; $\frac{1}{6}$; $\frac{3}{26}$

Step 3:

$\frac{2}{91}$; $\frac{1}{270}$; $\frac{3}{220}$; $\frac{2}{147}$; $\frac{1}{45}$; $\frac{1}{360}$; $\frac{1}{400}$; $\frac{6}{95}$

Test 18

Step 1:

3.6; 8.5; 3; 7.8; 9.5; 2.1; 3.5; 34.15

Step 2:

4.5; 7.5; 10; 0.5; 6.4; 1.8; 7.7; 9.2; 8.6; 6.4; 11.9; 2.19; 4.8; 1; 0.44; 4.8; 4.4; 7.8

Step 3:

0.67; 1.1; 13.5; 0.62; 0.46; 18; 17; 1.8; 18

Answers

Test 19
Step 1:
5.4; 3.25; 10.07; 1.8; 72.946; 2.21; 2; 90.315

Step 2:
14.08; 18.75; 89.09; 10; 18.1; 5.22; 27.3; 6.6; 44.55; 10.73; 21; 96.5; 17.2; 19; 25.5; 61.5; 9.78; 4.5

Step 3:
690; 9040; 4400; 490; 48; 1784; 7500; 18; 1.03; 4.6; 9608; 7, 560

Test 20
Step 1:
20; 4; 14; 5.4; 9.6; 2; 4.5; 2

Step 2:
1. 94.85; 16.49; 51.42; 27.64; 36.47; 88.12; 61.35; 9.26
2. 51.0; 18.0; 84.6; 17.2; 87.6; 71.2; 85.3; 13.3

Step 3:
1.5; 0.03; 0.012; 0.15; 0.32; 0.028; 1; 100

Test 21
Step 1:
0.8; 1.8; 1.02; 1.26; 1.92; 4.86; 4.32; 2.3

Step 2:
2.1; 0; 1.2; 0.2; 2.1; 0.4; 1.8; 0.57; 0.3; 2; 2; 1.22; 2.1; 1; 0.65

Step 3:
1.4; 0.6; 0.9; 0.1; 1.52; 2; 0.199; 4; 4

Test 22
Step 1:
1.1; 4.2; 1.3; 2.3; 2.3; 0.4; 1.8; 0.8

Step 2:
0.04; 0.02; 0.03; 2.5; 1.6; 5.2; 1.1; 2.1; 4.1; 0.09; 7.1; 0.16; 0.26; 0.12; 8.1; 3.4; 1.3; 3.2

Step 3:
1.1; 1.2; 1.3; 1.4; 0.16; 1.2; 1.3; 1.1

Test 23
Step 1:
0.7; 0.4; 0.08; 6.1; 0.1; 0.24; 2.3; 1.5; 0.03

Step 2:
0.5; 0.07; 0.3; 0.55; 0.11; 1.1; 0.14; 1.1; 0.6; 0.6; 0.07; 0.12; 0.08; 1.8; 0.12; 0.4; 2.5; 0.3

Step 3:
0.22; 0.41; 0.03; 2.1; 2.05; 0.21; 5; 0.2

Test 24
Step 1:
0.07; 0.4; 0.08; 6.1; 0.1; 0.04; 1.15; 0.5; 0.705

Step 2:
0.2; 0.05; 0.24; 0.08; 0.14; 0.37; 0.05; 0.85; 0.03; 0.18; 0.03; 0.024; 0.8; 0.09; 0.05; 0.002; 0.062; 1.21; 0.064; 0.002

Step 3:
1.403; 0.012; 0.011; 0.017; 0.11; 0.012; 0.15; 0.05; 0.003

Test 25
Step 1:
108; 7.5; 1.2; 0.9; 0.48; 0.3; 2.5; 0.4

Step 2:
0.4; 0.6; 2.4; 23; 12; 14.4; 0.8; 1.1; 0.13; 14; 0.7; 35.2; 6.5; 0.45; 25.6; 7.5; 4; 16

Step 3:
1.8; 1.6; 72; 130; 12; 20; 4.5; 2.4

Test 26
Step 1:
3; 0.9; 0.8; 0.2; 0.5; 3.2; 2.5; 16

Step 2:
5.6; 19; 27; 0.43; 24; 14; 0.6; 23; 1; 20; 0.8; 5; 0.5; 100; 20

Step 3:
35; 0.2; 0.2; 0.3; 4; 0.3; 20; 50

Test 27
Step 1:
1.4; 2.9; 4.2; 5.9; 19.4; 7; 42; 3; 12

Step 2:
23.4; 10.8; 9.8; 20.3; 16.7; 11.7; 7.5; 11.9; 2; 12; 12; 12.6; 6; 65; 740; 79.2

Step 3:
3.3; 12; 60; 22; 46.46; 49.5; 37.62; 4

Test 28
Equivalent answers are possible.

Step 1:
$\frac{1}{5}$; $\frac{951}{1000}$; $\frac{27}{100}$; $\frac{347}{1000}$; $8\frac{3}{5}$; $\frac{43}{100}$; $\frac{3}{50}$; $\frac{1}{125}$; $2\frac{9}{25}$; $5\frac{21}{1000}$

Step 2:
0.3; 0.48; 0.529; 0.075; 0.07; 2.1; 3.92; 2.408; 7.002; 1.064; 0.4; 0.375; 0.14; 0.16; 1.45

Step 3:
0.44; 0.625; 2.5; 0.56; 0.24; 2.25; 4.6; 2.25

Test 29
Equivalent answers are possible.

Step 1:
$1\frac{1}{6}$; 3.95; $\frac{1}{4}$; $\frac{1}{6}$; 1.925; 1.8; $\frac{1}{16}$; $\frac{1}{20}$

Answers

Step 2:

0.9; 3.125; 8.625; $\frac{3}{20}$; $\frac{1}{6}$; 5.95; $\frac{1}{120}$; 9.41; 1.26; $\frac{5}{7}$;

$4\frac{1}{4}$; $\frac{3}{152}$

Step 3:

39; 2.25; $\frac{5}{12}$; 75; $2\frac{3}{4}$; $2\frac{3}{11}$; $3\frac{1}{3}$; 37

Test 30

Equivalent answers are possible.

Step 1:

4.9; 11.55; $\frac{5}{108}$; $\frac{7}{24}$; 8.225; 4.15; $\frac{5}{224}$; $\frac{16}{27}$

Step 2:

0.8; 4.2; $\frac{9}{470}$; 9.78; $\frac{1}{9}$; $3\frac{1}{3}$; $4\frac{5}{12}$; $\frac{7}{184}$; $3\frac{1}{5}$; $\frac{1}{2}$; $8\frac{3}{4}$; $\frac{3}{80}$

Step 3:

164; 15; $\frac{7}{16}$; 70.8; 25; 13; 9; 32

Test 31

Step 1:

2*b*; 4*d*; 3*c*; 2*y*; 3*r*; 5*f*; 4*s*; 5*t*

Step 2:

a; 8*b*; 10*z*; 6*p*; 12*g*; *v*; *r*; 9*n*; 15*e*; 12*y*; 3; 6*t*; 3*b*; 27*a*; 4.5*b*; *x*

Step 3:

15*m*; 56*s*; 12*a*; 7.2*p*; 7.5*t*; 7*a*; 63*s*; 42*t*; 33*x*; 1.6*y*; 6.5*t*; 12*s*; 6*x*; 91*m*; 15*k*; 11*y*

Test 32

Step 1:

13*b*; 14*a*; 15*h*; 9*a*; *y*²; 108*c*; 15*c*; *b*; 7.5*x*; 2.2*x*

Step 2:

69*k*; 0; 54*m*; 31*d*; 48*s*; 120*t*; 4.56*a*; 34*y*; 7.5*p*; 9*q*; 60*x*; 10*c*

Step 3:

2*q*; *b*; 9*y*; 16*x*; 10*a*; 42*r*; 19*a*; 6*p*

Test 33

Step 1:

7; 8; 18; 15; 11; 1.9; 30.5; 9.7

Step 2:

18.2; 34; 17.5; 10.4; 200; 10; 1.8; 4.5; 36; 0; 70; 11.8; 10.8; 21.4; 15.9

Step 3:

4; 1.3; 3.1; 31.4; 70; 0.1

Test 34

Step 1:

6; 26; 19; 21; 33; 4; 2; 9

Step 2:

3; 2; 5; 10; 6; 1.3; 2; 0.3; 2; 0.5; 6; 10; 0.5; 8; 5

Step 3:

96; 24; 9; 31; 8; 1.6

Test 35

Step 1:

5:1;

2:5;

7:9;

6:7;

2:3;

3:5;

13:15;

4:7

Step 2:

7:4;

65:3;

1:5;

5:4;

2:3;

8:15;

9:2;

9:2;

2:1;

2:3;

1:40;

45:1

Step 3:

1:3;

8:1;

5:7;

15:2;

2:5;

1:20;

5:8;

4:3

Mind Gym:

4	1	6	2	7	3	5	8	9
9	7	5	4	8	1	2	3	6
3	2	8	5	9	6	1	7	4
6	5	4	1	2	8	3	9	7
1	8	3	7	5	9	6	4	2
2	9	7	3	6	4	8	5	1
8	3	2	6	4	7	9	1	5
5	4	1	9	3	2	7	6	8
7	6	9	8	1	5	4	2	3

Test 36

Step 1:

50%; 12.5%; 6.25%; 5%; 4%; 25%; 30%; 20%

Answers

Step 2:

40%; 37.5%; 31.25%; 45%; 370%; 16%; 225%; 160%; 35%; 8.7%; 780%; 12%; 540%; 100%; 20%; 1%

Step 3:

20; 80; 37.5; 35

Test 37

Step 1:

Number 2 only ticked.

Step 2:

$(26 + 30 + 27 + 29) \div 4 = 28$;

$(75 + 78 + 65 + 74) \div 4 = 73$;

$(18 + 18 + 24 + 25 + 20 + 21) \div 6 = 21$;

$(15 + 10 + 12 + 16 + 14) \div 5 = 13.4$;

$(29 + 25 + 23 + 21 + 20 + 26) \div 6 = 24$;

$(24 + 28 + 20 + 17 + 23) \div 5 = 22.4$

Step 3:

1. 80   **2.** 86.5   **3.** 44

Test 38

Step 1:

parallel lines; parallel to each other; perpendicular to each other; perpendicular line; perpendicular foot

Step 2:

First table, left to right: 154, 144, 210, 230, 156, 81.9, 13.2, 13.8, 16.5, 9.1

Second table, left to right: 154, 3, 2, 2, 0.5, 0.6, 0.3, 4, 3.5, 4

Step 3:

From left to right: 20, 20, 0.4, 16, 7, 1.9, 0.4, 0.4, 0.3

Test 39

Step 1:

From left to right: 120, 510, 25, 64, 135, 85, 750, 1250, 25

Step 2:

First table, left to right: 120, 66, 130, 250, 7.5, 24, 150, 1000

Second table, left to right: 810, 16.2, 3, 20.4, 14.5, 18.6, 9.6

Step 3:

First table, left to right: 72, 55, 12.4, 21.5, 19.6

Second table, left to right: 132, 31.2, 22, 19.5, 25

Test 40

Step 1:

$\frac{2}{33}$; $\frac{7}{54}$; $2\frac{1}{4}$; $\frac{11}{30}$; $\frac{1}{28}$; $3\frac{21}{44}$; $20\frac{11}{18}$; $\frac{3}{20}$

Equivalent answers are possible.

Step 2:

Numbers 1, 3, 5, 7, 8 and 9 ticked

Step 3:

1.
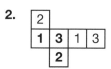

2.
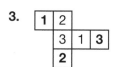

3.
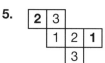

4.

5.
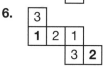

6.